The Rise and Fall of Animal Experimentation

The Rise and Fall of Animal Experimentation

Empathy, Science, and the Future of Research

RICHARD J. MILLER

OXFORD
UNIVERSITY PRESS

OXFORD
UNIVERSITY PRESS

Oxford University Press is a department of the University of Oxford. It furthers
the University's objective of excellence in research, scholarship, and education
by publishing worldwide. Oxford is a registered trade mark of Oxford University
Press in the UK and certain other countries.

Published in the United States of America by Oxford University Press
198 Madison Avenue, New York, NY 10016, United States of America.

Library of Congress Cataloging-in-Publication Data
Names: Miller, Richard J., 1950– author.
Title: The rise and fall of animal experimentation : empathy, science, and
the future of research / Richard J. Miller.
Description: New York, NY : Oxford University Press, [2023] |
Includes bibliographical references and index.
Identifiers: LCCN 2022055823 (print) | LCCN 2022055824 (ebook) |
ISBN 9780197665756 (hardback) | ISBN 9780197665770 (epub)
Subjects: LCSH: Animal experimentation.
Classification: LCC HV4915 .M55 2023 (print) | LCC HV4915 (ebook) |
DDC 179/.4—dc23/eng/20230130
LC record available at https://lccn.loc.gov/2022055823
LC ebook record available at https://lccn.loc.gov/2022055824

DOI: 10.1093/oso/9780197665756.001.0001

Printed by Sheridan Books, Inc., United States of America

For Marvel and Oba

"The greatness of a nation and its moral progress can be judged by the way its animals are treated" (*Mahatma Gandhi*)

Acknowledgments

I would like to offer my sincerest thanks to so many people who helped me write this book. First and foremost, my wife, Lauren, who was always available to read, correct, and offer suggestions. Other people who read sections of the book and offered their guidance included Anne-Marie Malfait, Rachel Miller, and my brother Daniel Miller. Many thanks to all of you.

I also really appreciate the advice I received from Betsy Lerner and my agent Christopher Rogers at Dunow, Carson and Lerner, who helped me navigate the world of book publishing—many thanks! Thanks also to my great editor Jeremy Lewis at Oxford University Press for all of his assistance. A big shoutout to Holly MacDonald and Matt Barraza, who helped me prepare the book manuscript. Thanks to you all!

Contents

1. The Seminar   1

2. Greek Awakenings   9

3. Circular Arguments   33

4. Mapping Humanity   67

5. Fear and Trembling   103

6. The Modern Prometheus   141

7. I Want to Be Your Dog   179

8. Not Just Kids   221

9. The Cloud Cap'd Towers   257

Bibliography   273
Index   277

1

The Seminar

I am sitting in the conference room listening to our departmental seminar. We have these most Monday afternoons during the academic year. Visiting scientists, members of the faculty, or students present their experimental data for an hour and answer questions about the research they are performing. Today is fairly typical. The speaker in this instance is a rather famous scientist from a prestigious university on the West Coast. He is very well known for the cutting-edge technology he brings to bear when investigating scientific questions. His lecture begins by defining a problem: it's to do with drug addiction, certainly an important issue these days. Why do people behave this way? What is the brain circuitry involved? What can we do about it? Rather than investigating these questions using human subjects, the speaker has decided to use an "animal model" of drug addiction. He thinks that mice can become addicted to drugs, and by studying mice, he will find out interesting things about humans, leading to the discovery of new cures for addiction. The use of animals to model human disease is an approach that many scientists use these days. After a general introduction, we are starting to get into the data portion of the lecture. The speaker is showing a video of a mouse. However, this isn't just an ordinary mouse. A mysterious black tower has been drilled into the animal's skull (Figure 1.1). The animal has numerous wires and tubes inserted into various parts of its body and into its brain, which are then fed into a variety of recording devices and computers. These machines can analyze how nerves in the brain respond during different types of behavior.[1] You can see exactly which brain circuits are activated in real time when the mouse is behaving in different ways.

Other wires inserted into the brain can deliver stimulating electrical pulses or flashes of laser light that activate specific nerves and initiate different types of behavior. On this particular occasion, we can see how the mouse's brain responds when it chooses to self-administer a dose of heroin. Next, we see how the brain looks when the speaker administers an "aversive" electrical pulse and how this alters the way the mouse chooses to self-administer the drug. The aversive stimulus causes the mouse to flinch. Other types of

The Rise and Fall of Animal Experimentation. Richard J. Miller, Oxford University Press. © Richard J. Miller 2023. DOI: 10.1093/oso/9780197665756.003.0001

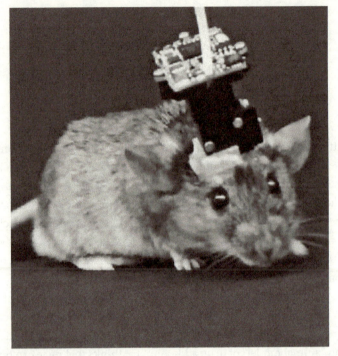

Figure 1.1 Recording device screwed into the skull of a mouse.[1]

stimulation cause the mouse to have a seizure. It rolls around on the ground, showing every sign of being in great distress. There can be no doubt about the aversiveness of the stimulus. There are numerous variations on this sequence of events. The speaker is talking about what is going on. "Here we see the application of a negative stimulus and the change in valence of the reinforcer," he says. The scientists in the audience all nod their heads in agreement.

What the speaker says certainly sounds very scientific. One half of my brain is intrigued by what he is saying. After all, I'm a trained scientist, and I have done experiments like this myself many times over the years. One can see why the speaker's research has garnered so much praise. His approach is ultra up to date in a technical sense, yielding data the likes of which have never been seen before. "That's certainly very interesting," I say to myself, and I can tell that my colleagues are similarly engaged with what the speaker is saying. I've been to hundreds of lectures like this, and I know whether a speaker has something interesting to say or not. However, there is another part of my brain that is viewing the scene with an entirely different set of emotions. Up on the screen we can see what is essentially a cyborg mouse.

There are more wires coming out of its brain than out of Boris Karlof in *Frankenstein*. Moreover, the aversive brain stimulation is precisely what it says it is. Frankly, anybody who wasn't a scientist would view the scene and conclude that they were observing an example of extreme cruelty to animals. There really isn't any other way of looking at it. In fact, the speaker clearly recognizes the situation. At some point, when he is about to test the effects of a drug on the mouse, he says, "Now that the mouse has gone through hell, let's see what this kinase inhibitor [a drug] does." It is clear, then, by his own admission, that the speaker acknowledges that the mouse is experiencing something extremely unpleasant. In other words, he understands that the mouse isn't just a machine but a conscious, living, breathing, and feeling organism. When the experiment is over the mouse is killed and its tissues are analyzed. But scientists never say the mouse has been killed. Instead, they usually use the word "sacrificed." But sacrificed for what purpose? Sacrificed to what god? For the audience, the death is a thing of beauty, a technological marvel they can appreciate. The mouse has been sacrificed to the god of technology. Like the protagonist in Kafka's story *In the Penal Colony*, the audience has participated in an appreciation of the aesthetics of death.

For the speaker this is not just a lecture—it's a performance; dazzling the audience is the name of the game. The game is about power, and it is being played for high stakes; the speaker's reputation is on the line. Like Cipolla, the magician in Thomas Mann's story *Mario and the Magician,* the speaker is not just trying to inform the audience but to dominate it. He is like a conjuror about to pull something out of his hat. Will it be a rabbit or something else? He has the audience on the edge of their seats. They want to know the answer. The speaker's performance and what it aims to achieve is part of a tradition that goes back to Roman times, to scientists like Galen. The speaker presents himself as a Renaissance magus inspiring wonder in his audience as scientists have always done. Really, his mouse-machine hybrids are nothing new. Giovanni Fontana's *Bellica Instrumenta* of 1420 described the design of a rabbit that could fart fire, and in Gaspar Schott's *Magia Universalis* of 1657 we can read about a cat piano in which a series of felines meowed out different notes when the piano keys were depressed, causing a pin to poke them. The fifteenth-century magus Regiomontanus is said to have made an iron fly and a wooden eagle, both of which were capable of flight. And in 1547, the great alchemist John Dee staged a version of *Pax*, Aristophanes's ancient Greek comedy, for which he created a gigantic dung beetle that, "flying up to the Jupiter's palace," caused "great wondering" and suspicion that it had been achieved by supernatural means. There are many other examples of the use of such "mathematical magic" to amaze an audience, and here at the seminar we have the latest example.

In this afternoon's seminar there are aspects of scientific experimentation involved in the performance, no doubt, but also aspects of competition, of being cleverer than everybody else through the use of the latest magical devices, of positioning oneself as the top dog in the field and reaping the rewards of celebrity and financial gain. The equipment the speaker employs for performing his experiments is very expensive—surely only top scientists can afford to do this type of research. This clearly includes the speaker. The funding agencies have picked him out as a star and have handed him ample funds for carrying out his studies. On this occasion everybody in the audience is in agreement that the presentation illustrates what is clearly excellent, cutting-edge science—but is it?

The idea that the events on display might also be cruel or unethical is not something that anybody in the room is thinking about. Afterward, during the discussion session, members of the audience comment on the speaker's data. Their choice of words is surprising. "These are elegant studies," says one member of the audience. Another declares that the precision of the data is "exquisite." This choice of words in the face of what could only be interpreted as extreme cruelty is interesting. One might find the same emotions expressed using similar terms in De Sade's *The 120 Days of Sodom,* in Octave Mirbeau's *The Torture Garden,* or perhaps in J. G. Ballard's *The Atrocity Exhibition,* or *Crash.* Here, torture and cruelty are appreciated purely for their aesthetic value—stripped of any emotional involvement. Scientists are taught that they must stand back and appreciate experiments, objectively free of the taint of emotional involvement. One can imagine Huysman's Duc Des Esseintes relishing this technological display as an exercise in aesthetics and thinking that it was anything but *Á Rebours* (Against Nature).[2] Of course, only we scientists can appreciate what we are seeing for its beauty because of our undoubted intellectual superiority and possession of relevant facts, of which most other individuals are ignorant and unaware. Here we may be misunderstood by the rest of humanity—again, according to Huysmans, "though the world may shudder at my joy, and in its coarseness know not what I mean."

So, what exactly is going on? Are the scientists in the audience simply a group of sadists? I know for certain that they are not. Indeed, I know the majority of them very well. Many of them have pet dogs and cats, which they talk about fondly. We swap Instagram pictures of our pets and gush over them like proud parents over their offspring: "Here is Doodles posing with her little toy—isn't she cute?" Somehow the scientists in question have been able to develop an extremely schizophrenic attitude toward the use of animals in their research. They have managed to put their experiments into a box where the normal rules of ethical behavior don't seem to apply anymore. At the very least, they are being extremely callous. But to quote Walter Benjamin on the matter, "Science (Art) is not the opposite of barbarity. Reason is not the contradiction of violence." The philosopher Martin Heidegger spoke of the process of "Gestell" or "enframing"—the ability of humans to enter into a technological way of thinking in which the world is seen as existing as a series of resources for human consumption, stripped of all other considerations. Heidegger expressed concern that people in the modern world were thinking about themselves, other people, and nature in general as things to be used (and used as efficiently as possible). This, of course, might be related to the general aims of capitalism.

Somewhere there is a disconnect that has been bothering me increasingly over recent years. When I first entered a laboratory as a graduate student desperate to play my role in contributing to the corpus of fundamental human knowledge, I had to kill rats to obtain brain tissue for carrying out my experiments on the effects of antipsychotic drugs on different neurotransmitter molecules in the brain. Killing rats in those days was an extremely hands-on affair. One was intimately involved with the entire process. First of all, one would pick the rat up by the tail and whack its head on the bench so as to stun it. Then their heads would be cut off with a "guillotine." When I first had to kill a rat, I found I just couldn't do it. After I made a rather feeble attempt at stunning the rat, it sat up on the laboratory bench, looked at me briefly, and then fled as quickly as possible down a hole in the wall that accommodated the building's heating pipes. Several other rats did the same thing, and their descendants may still be living there today as far as anybody knows. The first ten rats I tried to kill all escaped. My nickname in the lab was "the rat liberationist." Eventually, however, I "got the hang of it" and somehow managed to suppress the disgust I was feeling at what I was doing. After all, I was going to be a famous scientist and find out all kinds of really cool stuff about the brain. A few rats were just collateral damage. I decided not to think about it.

And, generally speaking, I didn't think about it. Except occasionally I did. But then I would always figure out a way of suppressing these kinds of thoughts. My scientific career was going along well. Soon I had my own laboratory, and my own students had the job of killing the animals while I just stayed in my office and was no longer confronted with the spectacle of bloody death on a routine basis. I had succeeded in "enframing" science, to use Heidegger's term. It was in a special box and played by its own rules. For reasons I will discuss later, mice are generally used these days rather than rats as the experimental mammal of choice. Mice are much smaller and easier to kill. Scientists use other methods today. Mice are first rendered unconscious by using CO_2 or an anesthetic prior to cervical dislocation. We never talk about "killing" animals. As indicated above, we say they are being "euthanized" or "sacrificed" or some other term to describe what we are doing. The idea that it is a sacrifice carries with it at some level an idea that the mice are participating willingly for our benefit—but one has to wonder.

Biomedical research as performed today clearly involves the institutionalized systematic abuse of animals. However, as scientists, we have concluded it is reasonable to behave in this way. As John P. Gluck puts it in a recent book,[3] "Although the practice of using animals as models for humans when direct experimentation on humans would be unethical is firmly entrenched in biomedicine and psychology, it rests on an almost hypocritical inconsistency. When researchers want to extrapolate their animal results to humans, the close biological similarities between animal subjects and humans are celebrated and emphasized. But when it comes to justifying doing to animals what can't be ethically done to humans, any similarities between animals and humans that might have ethical relevance—the experience of pain, fear and distress, the ability to suffer—are minimized, ignored, or denied."

What has made scientists indifferent to the suffering of the hundreds of millions of animals that are killed every year in the name of experimental research? In my experience, acceptance of this situation is so engrained in their psyche that most scientists have never even wondered about it. Justifying the use of animals in experiments is viewed as a waste of time: "Why do I have to bother with these things? It's just getting in the way of the science."

There are many questions that need to be answered. First of all, if your goal is to find out something about humans, why do experiments on animals anyway? When and where did this practice begin, and what useful information has it really yielded? Second, even if we can find out useful things by experimenting on animals, why would we do these things if they are ethically and morally abhorrent? Are there pressures on scientists today that make them do things that they may not want to do? What is it about our modern culture that brings these pressures to bear? And then there is the future. It is clear to me that advances in biology will make the use of animals in biomedical research obsolete in the near future; we are already racing toward this end. What are these technologies, and what hopes do they engender? In this book I want to consider some of these questions, even if I am unable to completely answer them. But perhaps if we scientists think about them, we may change our behavior accordingly.

One thing strikes me as I listen to the seminar. However technically proficient these studies may be, they have suddenly begun to look terribly old-fashioned. They simply aren't in tune with the times, either from a scientific or from a popular point of view. Our culture is changing. Humans are becoming more concerned with the fate of the world around them. Growing trends such as environmentalism, veganism, and animal welfare are the symptoms that herald humanity's concern about the fate of the planet we live in and the forces that are leading to its destruction. Can science respond to these issues in a way that allows it to contribute to the progress of civilization without encouraging barbarism? We need to think about how science can prosper in a world in which the voice of public opinion is starting to cry out for better treatment for all creatures and of the planet that nurtures us. Scientists have secluded themselves in their laboratories, cordoned themselves off so they don't have to account for what they do. Science must listen to the voice of the world at large and take what it says seriously. In the nineteenth century, the composer Richard Wagner spoke of the "Music of the Future." Clearly what we need to think about today are the "scientists of the future"; scientists who are not just investigators exploiting technology but who can account for their actions ethically. But how do we create the "scientists of the future?" In order to do so, we need to understand how science's current attitude toward animals developed; we need to understand the "scientists of the past." How did animal experimentation begin, and how did our modern approach to science come to pass? One may be surprised to know that science's current attitude toward animals is nothing new. To really understand it, we must turn our attention to events that occurred thousands of years ago in ancient Greece. It is there that our investigation will begin.

Notes

1. Daniel Aharoni et al., "All the Light That We Can See: A New Era in Miniaturized Microscopy," *Nature Methods* 16, no. 1 (January 2019): 11–13, doi:10.1038/s41592-018-0266-x.
2. J.-K. Huysmans, Robert Baldick, and Patrick McGuinness, *Against Nature*, Penguin Classics (London; New York: Penguin Books, 2003).
3. John P. Gluck, *Voracious Science & Vulnerable Animals: A Primate Scientist's Ethical Journey*, Animal Lives (Chicago: University of Chicago Press, 2016), 5.

2

Greek Awakenings

Clown
What is the opinion of Pythagoras concerning wild fowl?
Malvolio
That the soul of our grandam might haply inhabit a bird.
Clown
What thinkest thou of his opinion?
Malvolio
I think nobly of the soul, and no way approve his opinion.
Clown
Fare thee well. Remain thou still in darkness:
thou shalt hold the opinion of Pythagoras ere I will
allow of thy wits, and fear to kill a woodcock, lest
thou dispossess the soul of thy grandam. Fare thee well.

—*Twelfth Night*, William Shakespeare

Why does science, or at least science in the Western tradition, countenance the vivisection and killing of animals as an essential part of its modus operandi? I say "Western" because historically, in most other societies, the kind of scientific approach we are all familiar with today didn't exist until it was appropriated from Western culture. Nowadays, what we call biomedical research has similar characteristics all over the world and follows Western scientific practices, which have had spectacular success in helping us understand the laws of Nature. However, let us note, right from the outset, that there are other medical traditions, in great Asian countries like China and India, that never used animals as experimental subjects. These traditions are extremely old and have often been very effective. The ancient Chinese discovered many extremely valuable therapeutic agents that are still used today. These discoveries resulted from observing the effects of natural plant

The Rise and Fall of Animal Experimentation. Richard J. Miller, Oxford University Press. © Richard J. Miller 2023.
DOI: 10.1093/oso/9780197665756.003.0002

products directly on humans. No animal experimentation was necessary. Consider the important antimalarial drug artemisinin, identified in traditional Chinese medical texts as a component of the Sweet Wormwood plant (*Artemesia annua*). It was then rediscovered in modern times by the Chinese scientist Tu Youyou, research for which she became the first Chinese woman to win the Nobel Prize. Indeed, many of the most important drugs we use today have origins that are both ancient and originally derived directly from Nature. They were discovered by humans through self-experimentation prior to the time when modern science developed in the seventeenth century. These include drugs such as digoxin, quinine, morphine, the salicylates, cannabis, and countless others. Of course, in ancient times people had little idea as to exactly how these substances produced their beneficial effects. Sophisticated explanations describing their mechanisms of action would have to wait until modern Western science provided the intellectual context for insights of this type.

Based on our knowledge about how therapeutic agents and procedures have been developed throughout human history, one may well wonder whether killing animals is really inextricably linked to the progress of the biological sciences and medicine. If you are an active participant in most types of biomedical research these days, it is usually assumed that this is the case and, for most practicing scientists, it seems that it has always been this way. The animals employed in contemporary research may be as lowly as worms and flies; they may be fish, or they may be mammals—frequently rats and mice but also rabbits, cats, dogs, pigs, sheep, and monkeys. Scientists in the field of biomedical research are taught that using animals in their research is absolutely necessary. Hence, for science to proceed in the way it does, scientists need to develop a particular attitude toward animals, that it is reasonable to perform experiments on them, things that we absolutely would not do to one another, and, moreover, that these experiments will reliably yield important information that is relevant to solving problems that afflict the human race. But for what reason and how did this attitude develop in Western countries when it certainly wasn't the case in other parts of the world? Answering this question is not straightforward and requires us to reach back to ancient times when the enterprise we call science had its beginnings.

Animal experimentation is not a modern invention. Of course, human beings don't just use animals for scientific research. We also kill animals to provide food and clothing, and, in ancient times, animals were frequently used as sacrifices to please the gods or as forms of entertainment where

they were killed in arenas in front of large appreciative audiences, something that is regrettably still going on today—consider bullfighting, for example. Many human beings are of the opinion that animals can be used for
any purpose that they think will serve humanity. It is a one-way street. It is
considered acceptable for us to eat animals, but it isn't acceptable for animals to eat us. Animals are not our equals. Just like the rest of our planet,
animals are viewed as resources; they are here for us to exploit for our own
good. And that is precisely what humans have done very effectively over the
centuries. Animal experimentation carried out by humans is another aspect
of our anthropocentric attitude, the idea that the planet Earth and everything
it contains is available to serve human interests. Of course, human beings are
entitled to act in ways that ensure their general health and security and to
attempt to solve problems that negatively impact their lives. But surely, what
we do must be considered carefully and must take place within some kind
of ethical boundary. Otherwise, we are in danger of compromising the core
of our humanity and, from a practical point of view, destroying the planet
that supports our existence. Moreover, is it really the case that animal experimentation carried out by today's scientists reliably yields important scientific
information? I consider that a discussion of the real usefulness and ethical appropriateness of animal experimentation has many important implications
for the way in which human beings live their lives in today's world.

The Dawn of Science

One thing we know is that animal experimentation for scientific purposes
began around 2,500 years ago in ancient Greece, together with the development of procedures that we would recognize, even today, as constituting
experimental biomedical research. What do we mean by science? Science
believes that there are laws that govern what happens in the universe—that
things don't just happen by chance. According to one definition, science
constitutes the "orderly and systematic comprehension, description and/or
explanation of natural phenomena"[1] or, perhaps tellingly, "the system of behavior by which man acquires mastery of his environment."[2] Science helps
us not only to describe our world accurately but also, most importantly, to
derive laws through which we can predict the future and become the masters
of Nature. The enablers of this mastery are the scientists. They are the people
whose job it is to bend Nature to the will of humankind, forcing her to do

what we desire. The Abrahamic religious tradition has taught us, whether we are conscious of this influence or not, that humans are the pinnacle of all existence. We have been "chosen" by the Almighty and have been created in God's image, and the universe and everything in it are here to serve our needs. Science is the most powerful tool we have at our disposal for achieving this end.

Ancient representations of animals in cave paintings or sculptures, some going back as far as 40,000 BCE, indicate that humans didn't always think of animals in an "us and them" manner, purely as things they could use to their advantage. The worlds of humans and animals originally seemed to be much more of a shape-shifting continuum where the two influenced one another in all spheres of existence. Indeed, animals exhibited powers that sometimes made them superior to humans, and it was thought that these powers might be transmitted to humans through magical procedures resulting in powerful hybrid creatures. Animals represented what anthropologists call a "totem," and as the anthropologist Claude Levi-Strauss wrote, animals are not just "good to eat" but also "good to think." The barriers between what was human and what was animal were extremely porous, and one might easily pass from one domain to another allowing humans to take on the characteristics and powers of different creatures. But, according to one widely discussed historical model, beginning around 10,000 BCE with the Neolithic revolution, humans began to develop stable communities and agriculture together with an increasingly anthropocentric attitude that viewed the entire planet, including its animals, merely as resources to be exploited. However, to exploit animals for scientific purposes, first we had to invent science.

The cultural endeavor that we call science, in the form that we would generally recognize it today, had its origins in ancient Greece. During the period between 800 BCE and the fall of the Roman Empire in 476 CE, biomedical research developed many of its defining characteristics, including the use of animals as research tools. In prescientific times, the earliest Greek writings illustrated an Edenic society in which humans and animals lived a life of mutual respect. Hesiod writing in his "Works and Days" in the eighth century BCE described a Golden Age, an era of peace and prosperity, during which time people and animals lived peacefully together. Of course, we also find similar stories in Genesis and in contemporary ancient writings from Asia. Earth had enough vegetation to feed everybody and so animals, including humans, didn't have to kill each other for food or any other purpose. During this time the titan Cronos ruled the heavens and Earth. With the fall of Cronos and the rise of his son Zeus, Hesiod portrayed the coming of several subsequent ages: the Silver, the Bronze, and the one in which he lived, the Iron Age. With the end of the Golden Age came periods of increasing greed and violence. In these later ages, Hesiod wrote that humans began to eat the flesh of other animals and to exploit them in different ways. Hesiod also wrote that it was Zeus who began to differentiate between man and animals when he gave humans justice, a gift not shared with other creatures. This is a key point because you are not going to kill and eat an animal that you feel is morally your equal and with which you have an important spiritual link of some kind. However, from sometime around this point in history, ideas began to develop in ancient Greece that humans were "superior" to other species, suggesting that it was morally reasonable to exploit them, a principle that we will see became further enshrined in the writings of Aristotle and his followers and continued right through to the rise of Christianity and beyond to the present day. It was this emerging view of humans and animals as "us and them," rather than a continuum, that provided the important rationale for the experimental use of animals by humans. It wasn't until the nineteenth century that figures like Darwin showed how humans and animals were essentially connected in terms of their evolutionary development, by which time human attitudes toward animals had already crystallized into their master and slave characteristics.

And it wasn't only the principle of justice that the ancients thought differentiated man from the other animals. For example, one of Aesop's fables, written around 600 BCE, relates that animals angered Zeus by complaining about his rules. As a result, Zeus took language away from animals and gave it to man. The inability of animals to use language has frequently been used as an argument, since ancient Greek times until the present day, to suggest that animals cannot reason, and, if they cannot reason, that they are inferior to humans. This ancient idea that animals can be discriminated against owing to their supposed lack of reason is something that has also resonated down through the ages up until modern times. It is still something that is actively discussed today, particularly in view of recent scientific revelations that animals clearly use highly sophisticated forms of communication. Whether this really constitutes language in the human sense of the word or not is debatable, but perhaps that doesn't really matter, and as noted by the eighteenth-century naturalist Gilbert White, "The language of birds is very ancient, and, like other ancient modes of speech, very elliptical; little is said, but much is meant and understood."[3]

One of the great and defining achievements of the Greeks was their interest in seeking explanations for natural phenomena that went beyond the merely divine—a move from "mythos" to "logos." There are several different ways of obtaining reliable scientific knowledge. The first of these, which must have been routinely used by ancient man, is what we would call the study of natural history. There is much to be learned by simply looking at animals, plants, and the natural world in general and, if one does so in a systematic way, recognizing patterns in how Nature goes about its business. As the great seventeenth-century writer on science Sir Thomas Browne put it, "What Reason may not go to School to the wisdom of Bees, Ants, Spiders? Ruder heads stand amazed at the prodigious pieces of Nature, Whales, Elephants, Dromedaries, and Camels; these, I confess are the Colossus and majestick pieces of Nature: but in these narrow engines there is more curious Mathematicks; and the civility of these little Citizens more neatly sets forth the Wisdom of their Maker."[4] Natural history involves making observations of plants and animals in their normal "natural" environments without trying to interfere with them in any way. Such a procedure can be seen laid out clearly in certain books such as Gilbert White's classic *The Natural History of Selbourne* published in 1789, in which an English squire recorded his observations of the world around him. By carefully noting the characteristics of bird migration, the behavior of animals, the weather, and other natural phenomena, White could begin to recognize patterns that might be used to decipher the laws of Nature. It is this ability to understand patterns in the structure of Nature and use them to predict future outcomes that constitutes the essence of science. The kinds of observations made by Gilbert White were the precursors of modern disciplines like ecology. On the other hand, the practice of modern biomedical research often involves performing experiments, and most of us, even if we aren't scientists, have some idea as to what that entails. Usually an experiment, in the modern sense of the word, involves not only observing a phenomenon but also employing some artificial intervention to see how the behavior of what one is observing changes; an experiment, then, is a set of artificial procedures designed to help us understand how something in Nature works.

Modern research in the biomedical sciences frequently employs animals as experimental subjects. When and why did systematic experimentation on animals begin? If we want to find out things about humans, why don't we just perform experiments on humans? Humans and animals are obviously different in many respects, so why would experimentation on animals be useful?

The fact that we would use an animal in an experiment has several important implications, particularly if the experiment involves doing harm to the animal, which it almost invariably does. For one thing, it demands that humans have an attitude that allows them to perform experimental procedures on animals when we wouldn't do the same things to other humans. We would have to develop a way of thinking in which humans and animals were absolutely distinguished in fundamental ways so that different rules would apply to "us" and "them."

As we have seen, it was already clear from Hesiod that some consideration had been given to the existence of fundamental differences between humans and animals from the earliest times in Greece. As a reflection of this, in the fifth century BCE, one of the earliest Greek biologists, Alcmaeon of Croton, wrote that "man differs from the other creatures in that he alone has understanding, while other creatures have perception but do not have understanding." As we shall see, this kind of thinking would subsequently develop into the different types of arguments about animal intelligence and the moral standing of animals that persist to this day.

On the other hand, an absolute distinction between humans and animals wasn't the only kind of idea that developed in ancient Greece on this topic. Some early Greek ideas, such as those of Pythagoras and his followers in the sixth century BCE, considered that there were very important and permeable links between all living creatures; the souls of humans and animals could be transferred to one another horizontally, a process known and metempsychosis. According to legend, one day Pythagoras came upon somebody beating a dog. He urgently requested him to stop, saying that he recognized the dog's barking as the voice of one of his deceased friends! Not surprisingly, Pythagoras and his followers were strict vegetarians. The idea of the transmigration of souls was also described in Plato's *Republic*, which ends with a recounting of the Myth of Er. This story vividly relates the experiences of Er, son of Armenius, who dies but is reborn several days later while on his funeral pyre. Er tells his audience that during this time he had visited the afterlife and had returned to tell humanity what happens and how the ultimate fate of a human soul is decided. It turns out that because the soul is immortal, it isn't destroyed but can be transformed by metempsychosis into another being, which might be a human or an animal. As Er relates, because of the problems that afflict humanity, many philosophically sophisticated individuals prefer to be reincarnated as animals. Orpheus, for example, becomes a swan, and Agamemnon becomes an eagle. Following their informed choice, all individuals (except Er) have to drink the waters of the river Lethe, which produces complete forgetfulness as they begin their new lives. However, in spite of opinions of this type in the ancient battle of ideas, other attitudes toward animals ultimately prevailed when it came to the use of animals by science.

Probably one of the major influences behind the development of animal experimentation was an increasing desire to understand how we can improve human health, and this remains true when we look at the reasons that biomedical research is funded by governments today. In antiquity, biomedical science with clearly identifiable modern characteristics really began with the work of Hippocrates and his colleagues in the fifth century BCE. This involved the development of the kind of thinking that could provide scientific as opposed to divine explanations for disease. Indeed, around this time an important and influential theory was developed that allowed for the systematic treatment of disease. The theory stemmed from the belief, originating with Empedocles, a follower of Pythagoras, that everything was made from four elements—fire, air, earth, and water. Food and drink, which nourished the bodies of humans and animals, were made up of these four elements, which, upon digestion and respiration, were transformed into specific bodily fluids, known as the four humors—yellow bile, black bile, phlegm, and blood. The four humors were also associated with other attributes. Blood was hot and wet; phlegm was cold and wet; yellow bile was hot and dry; and black bile was cold and dry. Health consisted of maintaining humoral equilibrium in an individual at least within reasonable limits. Illness resulted when an excess or a deficiency occurred in one or more of the humors, producing a state of being "out of balance," or *dyscrasia*.

Of course, the fact that disease had a humoral basis rather than being divine also implied that humans could intervene and do something about it by carrying out procedures that would help the humors resume a state of *eucrasia*, or balance. The physician's task was to diagnose which humor was out of balance. Treatment then focused on restoring equilibrium by changing one's diet or by reducing the offending, out-of-balance humor by draining it out of the body using bloodletting, purging, or some other method. Even today, of course, we frequently speak of our health as a matter of "balance," an echo of the humoral theories of prior times.

Hippocrates's scientific attitude to disease can be seen from an example taken from his thesis *On the Sacred Disease*, which described the symptoms and treatment of epilepsy. Epilepsy was called the "sacred disease" because, in antiquity, it was thought that the symptoms were inflicted on humans by the gods as a punishment. Hippocrates didn't have much time for this kind of thinking. He wrote, "I do not believe that the 'sacred disease' is any more divine or sacred than any other disease but, on the contrary, has specific characteristics and a definite cause. Nevertheless, because it is completely different from other diseases, it has been regarded as a divine visitation by those who, being only human, view it with ignorance and astonishment." Even though Hippocrates's primary method was to obtain information about diseases by observation of human symptomology, he also suggested that if you want to find out what is going on, you might dissect an animal; in the case of epilepsy, he chose a goat that had died as the result of seizures. Hippocrates reported, "If you cut open the head, you will find that the brain is wet, full of fluid and foul smelling, convincing proof that disease and not the deity is harming the body." As one can see from this example, even in ancient times, people were beginning to think of animals as being similar enough to humans in some respects that they might be used as experimental stand-ins for human beings. It seems inevitable that as time went on, people would come to realize that when they looked inside a dead animal, the general shape of its organs did indeed often resemble those found in humans.

But why assume that what you find in an animal has any relevance to what is going on in a human? Why not dissect the brain of a human who had died of epilepsy, as we might do today at a postmortem examination? Wouldn't that be a much more straightforward and informative thing to do? As it turns out, there was a very good reason for not doing this. In ancient Greece and Rome, dissecting a human corpse was something you just didn't do because a corpse was considered to be a potent source of pollution. It was thought that anybody who came in contact with a dead body would be contaminated by it, and there were extremely strict laws about how a corpse should be handled. The lack of availability of human material for routine investigation is certainly one important reason people turned to the use of animals. Nevertheless, it is clear from reading the works of Hippocrates that, despite the ban on dissecting human corpses, there was a good deal of information available about the anatomy of human organs such as the lungs, liver, and heart. Some of this knowledge presumably came from observation of soldiers who had had deep wounds inflicted on them in battle as well as possibly from aborted fetuses. But it is also clear, as in the example above, that animal dissection was sometimes carried out with a view to understanding human disease. In other words, it was beginning to dawn on people that the anatomical similarities that existed between animals and humans could be used to argue by analogy that something that was observed in the former might also be true of the latter.

The Great Chain of Being

A detailed theoretical basis for the use of animals as experimental subjects in science first resulted from the thinking of the great Greek philosopher Aristotle, who was born in 384 BCE at Stagirus, a Greek seaport on the coast of Thrace. Aristotle's teacher Plato had developed a completely theoretical attitude toward science. On the other hand, Aristotle thought that a practical material approach, one in which he engaged with physical objects such as animals so as to obtain primary data, was the best way to proceed. As he went about his investigations systematically, Aristotle began inventing the experiment-based research program that most biomedical researchers carry out today. His attitude, that it is important to know about things for their own sake, but that these things may ultimately be useful for practical human concerns, is the credo that still drives most contemporary biomedical research.

Aristotle's approach was informed by the concept of teleology, which dictates that all things are the way they are for a specific purpose, something that is related to what we nowadays call "intelligent design." Hence, Aristotle believed that the best way to comprehend the universe was to understand the purpose for which everything was designed. For example, when thinking about how snakes copulate, Aristotle wrote in *On the Generation of Animals*: "Snakes copulate by twining around one another; and they have no testicles and no penis, as I have already observed—no penis because they have no legs . . . no testicles because of their length"—everything fits together according to a divine plan. Ultimately Aristotle thought that the purpose for which humans were designed, that is, our "final cause," was to fulfill our role as rational beings, something that made us unique and differentiated us from other animals. And how exactly were we different? In his treatise *On the Soul*, Aristotle wrote that the animal soul consisted of two main parts—a nutritive part, which governed things like metabolism and reproduction, and a sensory part, which contained their vital essence. In addition, the human soul included a rational part that set humans apart from all other animals. Here Aristotle was distinguishing between humans and animals vertically rather than considering them related horizontally by a process such as metempsychosis as described by Pythagoras and Plato. Because of the differences that Aristotle determined separated humans from nonhuman animals, he also thought that it was reasonable for humans to use animals (or plants, for that matter) for any purpose that might be of service to them. Specifically, in *The Politics*, he wrote, "In a like manner we may infer that, after the birth of animals, plants exist for their sake, and that other animals exist for the sake of humans, the domesticated for their use and food, and the wild ones, if not all of them, then at least the majority of them, for their food and clothing and various tools. Now if nature makes nothing incomplete or in vain, the inference must be that all animals exist for the sake of humans." According to Aristotle, everything had its place in the "great chain of being," where the only things that existed above humans were the gods and heavenly bodies. Aristotle's views about animals were complicated and nuanced. He certainly did attribute many qualities to animals, including a degree of intelligence. Nevertheless, it was clear that their abilities were inferior and often qualitatively different from those of humans.

Aristotle considered the investigation of animals to be central to the theoretical study of Nature. In his thesis *On the Parts of Animals,* Aristotle wrote that observation of the external appearance of animals was not sufficient for many purposes and needed to be supplemented using dissection. In *The Enquiry Concerning Animals*, he wrote, "The inner parts of man are mostly unknown and so we must refer to the parts of other animals which those of man resemble and examine them." He also wrote, "Ought we, for instance (to give an illustration of what I mean), to begin by discussing each separate species-man, lion, ox, and the like-taking each kind in hand independently of the rest, or ought we rather to deal first with the attributes which they have in common in virtue of some common element of their nature, and proceed from this as a basis for the consideration of them separately?" In other words, we could dissect animals to see how their organs look and, by analogy, infer the function of human organs.

A point of fundamental importance when considering Aristotle's scientific program is that his use of animals depended on his ability to distinguish between animals and humans by virtue of the properties of their souls. In other words, the world of animals and humans was not one of continuity but one of fundamental difference. Aristotle argued that because animals didn't possess a rational soul, they were devoid of moral status. Aristotle clearly loved animals, but it seems that he thought that there could be no such thing as "justice" when applied to animals as it was to humans. This is, of course, extremely important for any discussion as to the appropriateness of using animals as research tools because, if it could be demonstrated that humans and animals did not differ in this fundamental respect, it would open the door to a consideration that animals had rights that were similar to those of humans, as some contemporary thinkers have argued. Since antiquity, a sharp dichotomy between animal and humankind has been posited by those who, for one reason or another, are eager to claim a unique and privileged position for humanity in the spectrum of creation. Many aspects of this conversation are still the same today. All one has to do is change the word "soul" for a word such as "consciousness" and the character of the debate is virtually identical today to that of ancient times. Nowadays, however, there is also a wealth of scientific information that we can consider and that can help us to understand the higher mental functions of animals and their similarities, or otherwise, to humans.

Aristotle's views on the moral status of animals weren't the only ones discussed in antiquity. Just as today, it is clear that a wide variety of opinions existed on this matter. As we have already seen, prior thinkers like Pythagoras didn't think about the status of animals in the same way as Aristotle. In the years following Aristotle's death, many other writers commented on his views and what they meant. Some agreed with his attitude toward animals, and others didn't. For example, Aristotle's most famous student, the great biologist Theophrastus, took issue with Aristotle's thinking on the subject. Rather than harping on the differences between man and animals, Theophrastus accentuated their similarities. Animals had different abilities than humans, but their soul was part of the same spectrum as the human soul, and this relationship between animals and humans was why Theophrastus was against animal sacrifice. In his treatise On Piety, Theophrastus argued that animal sacrifice was an unjust act and not compatible with a human view of what constitutes holiness.

Following Aristotle and Theophrastus, the discussion about how to treat animals continued throughout antiquity. On the one side, there were those who followed Aristotle's lead, particularly the school of thinkers known as the Stoics, whereas others such as Plutarch and the Neo-Platonist Porphyry followed Theophrastus. According to Stoic teaching, the irrationality of animals was a consequence of the defective nature of the animal soul. Cicero, reviewing the opinions of the Stoics with respect to animals, wrote that animals were without reason and speech and so humans could use them in any way for their advantage without injustice. As we have seen, the fact that animals lack language has been used time and again since antiquity as a fundamental argument as to why they cannot be truly rational. Plutarch, on the other hand, writing in the first century CE in his *De sollenia animalium* (*On the Cleverness of Animals*), refuted the Stoic position, providing examples of animal behavior that seemed to argue in favor of their rationality. He was completely against eating animals, writing, "But for the sake of some little mouthful of flesh, we deprive a soul of the sun and light and of that proportion of life and time it had been born into the world to enjoy." Similar arguments were provided by the Neo-Platonist philosopher Porphyry writing in the third century CE in his *On Abstinence from Animal Food*—that animals are rational like humans, making it unjust to kill them for food or other purposes. It is therefore quite clear that Aristotle's way of conceptualizing the cognitive life of animals was only one of many opinions discussed in antiquity.

Nevertheless, it was Aristotle's experimental scientific program, including his use of animal dissection, that became the major influence on the development of science. Sometimes the acceptability or otherwise of such experimental procedures took an extremely surprising turn. One of the most remarkable occurred in Alexandria in the early third century BCE, with the introduction of the systematic experimental dissection and vivisection of that unique animal—the human being—as practiced by the Greek physiologist Herophilus and his younger colleague Erasistratus.[5] This was particularly surprising given the fact, as we have discussed, that there were incredibly strict rules in ancient Greece concerning the treatment of the human body. At the time, Alexandria was ruled by the Ptolemaic dynasty, the Greek successors to the Egyptian pharaohs, who had considerable ambitions for making the city a great center of learning and culture. Alexandria was an interesting mix of Hellenic and Egyptian influences with an extremely cosmopolitan population, and it attracted many outstanding scientists and intellectuals. The Ptolemies appear to have extended generous patronage to artists and scientists alike and were very open-minded about the kinds of practices they encouraged. Incredible as it may seem, it appears that they gave orders that condemned criminals could be handed over to Herophilus and Erasistratus for anatomical and even vivisectory experimentation. As can be imagined, the availability of tissues obtained from humans that could be studied in situ, while individuals were still alive, allowed Herophilus and Erasistratus to make numerous important discoveries that were directly relevant to humans and to the development of medicine. For example, they succeeded in distinguishing between the ventricles of the brain, discovered the nerves as unique entities separating nerves from blood vessels, provided a description of at least seven pairs of cranial nerves, and distinguished between sensory and motor nerves. Unfortunately, most of their original writings are lost to us, perhaps being destroyed by the fire that is known to have consumed the great Library of Alexandria where many scholarly volumes were kept. Most of what we know about these two scientists has come to us from the subsequent writings of people like Galen. One will perhaps not be surprised to learn that some commentators in antiquity, such as the Roman doctor Celsus, writing in the first century CE, declared that despite the advances they made, the human vivisection carried out by Herophilus and Erasistratus was inhumane and unethical, something I think most people would agree with today. Nevertheless, it does raise the issue, if we extend these arguments to thinking about animal experimentation in general: are scientific gains the only things that matter when we consider experiments that use animals?

Galen the Great

The experiments that Herophilus and Erasistratus carried out on humans were certainly not forgotten, especially by the man who ultimately proved to be one of the most influential biomedical researchers in history, particularly through his use of animal experimentation. His name was Aelius Galen, and he was born in 129 CE into a prosperous family in Pergamum, a culturally Greek city in what would nowadays be the northwestern part of Turkey. As a young man, he became very ill but recovered and was therefore encouraged to devote himself to Asclepius, the God of medicine, and to become a doctor. Galen traveled a great deal in his youth and studied in Smyrna, Corinth, and, significantly, Alexandria, where he became acquainted with the works of Herophilus and Erasistratus. This extended period of study lasted for a dozen years before he returned to Pergamum in 157 CE, where he was appointed to the highly competitive post of surgeon to the gladiators—something that speaks to his early promise as a doctor. This was the next best thing to being allowed to perform human dissection because, naturally, the job of a gladiator is to "dissect" his opponents in competition, and gladiators frequently killed or inflicted terrible wounds on one another. This might be considered, by a doctor at any rate, as a wonderful opportunity for increasing one's knowledge of human anatomy and physiology. Galen took full advantage of the situation. Although, as we shall see, he became one of history's greatest experts in animal dissection, he never considered this to be a substitute for what you could learn from studying humans and regularly sent his students to Alexandria to view human skeletons that remained there from the time of Herophilus and Erasistratus.

Galen was certainly very ambitious; in 162 CE he traveled to Rome, the center of the empire, to try and make a name for himself. He wasn't the only one to do this. Recent studies involving the sequencing of genomes from human remains buried in Rome over a long period of time have shown that in the first couple of centuries CE there was a huge influx of people from the Eastern Mediterranean into Rome,[6] presumably attracted there because of opportunities for professional advancement. Galen already had an excellent reputation and soon became an important fixture in the Roman professional medical milieu. Indeed, in 168 CE, he became doctor to the emperor Marcus Aurelius and his son Commodus and held this position at the Roman court until he died in 210 CE. It is clear that Galen wasn't the only doctor working in Rome at the time who had a highly successful and lucrative career, but as far as history is concerned, he was by far the most significant. There are several reasons for this. One is that he was certainly a brilliant doctor and scientist. However, he was also a philosopher and voluminous writer. Much of his writing was preserved in one form or another and so proved to be extremely influential over the centuries, particularly during the period of time following the fall of the Roman Empire when there was little further scientific progress for over 1,000 years. Moreover, the sheer scope of Galen's writings, what has become known as the "Galenic Corpus," was extraordinary. Galen produced more written work than just about any author in antiquity, including hundreds of books and treatises. Interestingly, Galen wrote completely in Greek rather than Latin. Indeed, there is very little evidence that Galen ever even learned to speak Latin well.

Galen was greatly influenced by the work of Hippocrates and the theory of the four humors. As some have commented about the history of medicine, "Hippocrates was God and Galen was his prophet." Other important influences on Galen were Aristotle and his followers the Stoics, as well as the work of Herophilus and Erasistratus. Galen made extensive use of animals as a substitute for humans because the dissection of humans, either alive or dead, was forbidden once again in Galen's time. Indeed, it seems that Galen never actually dissected a human corpse. Data on humans could be taken directly from the historical work of Herophilus and Erasistratus, from opportunities resulting from the examination of severely wounded patients such as gladiators, or perhaps even from the study of slaves, who were regarded as human but were also entirely dispensable and so the rules were sometimes bent with respect their treatment.

The prevailing ban on human dissection made Galen feel that performing experiments on animals was absolutely necessary, and with Galen, the extent of animal experimentation reached new heights.[7] Even though we know, with the benefit of hindsight, that Galen's scientific conclusions were not always correct, his achievements in the field of comparative anatomy and medicine, as well as his subsequent influence, cannot be denied. The scope of his animal experimentation and its incredible cruelty presumably required him to develop some type of rationale that allowed him to carry out his procedures with a clear conscience. His ideas harkened back to those of Aristotle and the Stoics placing animals below humans in the "great chain of being." By working within this philosophy and using it to his advantage, Galen built up his methods and discoveries around Stoic ideals. Animals were characterized as being fundamentally different from humans because of the superiority of the human faculty for reason. Galen certainly embraced the idea of the nonrationality of animals and used it liberally in the descriptions of his experiments. He wrote, "It is surely more likely that a non-rational brute, being less sensitive than a human being, will suffer nothing from such a wound." What is really amazing (or troubling) is that you might hear more or less the same thing said by many scientists today.

Galen operated under the belief that there were important analogies between animals and humans, but he was very selective about it; he thought that there were physical but not emotional similarities between them. Galen's rejection of sophisticated animal sentience allowed him to use animals closest to man such as apes because of their physical attributes, without concerning himself with their subjective experiences during his experiments. His opinions allowed for the vivisection of animals but also created a contradiction. If animals like apes were close enough to serve as reliable analogs to human anatomy and physiology, might there not also be similarities in their abilities to reason and to feel? But Galen never made such a connection. Here again, this contradiction still echoes down the ages to our contemporary use of animals in scientific research, and arguments about the actual status of animal sentience and consciousness remain a controversial, but critical, area of debate (see Chapter 5). Galen dissected numerous dogs, pigs, monkeys, apes, bears, and even, on a couple of occasions, elephants. His writings contain detailed anatomical descriptions of these animals and step-by-step explanations of the experimental procedures he performed on them. It is clear with Galen that we see, for the first time in history, an extensive program of animal vivisection that resembles how scientists investigate biomedical problems today.

Another aspect of Galen's approach that also resonates with contemporary science was its competitive nature. Galen was in a position where it was necessary for him to demonstrate that a procedure that he had developed was superior to that employed by any of his rivals. It wasn't just important for Galen to involve himself in experimentation; he had to be seen doing it. Galen's dissections weren't just scientific procedures; they were also, more often than not, actual theatrical performances, which took place in front of large and appreciative audiences in the courtyards of the houses of wealthy, influential citizens. Galen wrote that the audiences might also contain several colleagues whom he wanted to impress with his skill. As well as performing specific anatomical procedures, Galen would consider requests from the audience designed to demonstrate his virtuosity. These considerations can be compared to the situation described in the previous chapter where scientists give lectures at a meeting or to an academic department today to demonstrate their experimental skills and intellectual achievements.

It is clear from his voluminous writings that Galen conducted thousands of animal dissections and vivisections. Freed from the necessity of any moral or ethical consideration for his animal victims, Galen could literally do anything he pleased—and do it without recourse to an anesthetic or other intervention that might have provided his victim with some relief. Indeed, this was often precisely the point. For example, one of Galen's anatomical discoveries was the nature of the nerves that controlled the larynx—"the finest nerves, a pair of them like hairs," and so controlled the ability of an animal to "speak." For his demonstrations, Galen liked to use pigs or goats because he deemed these animals to have the loudest voices and as a result they would scream in pain the loudest when he cut into them. Then he would sever the laryngeal nerves, and the unfortunate animal would be deprived of its ability to even cry out. Sometimes Galen didn't agree with his illustrious antecedents such as Aristotle as, for example, when it came to the importance of the brain as the seat of intelligence—Aristotle thought it was the heart. To demonstrate the importance of the brain, Galen would perform experiments on live animals that involved removing parts of the skull and meninges (apparently accompanied by torrents of blood) and then manipulating parts of the brain. For example, an animal could be blinded by pressing on the anterior ventricles in the region of the optic nerve root.

Despite his incredible cruelty, it is certainly true that Galen used the results of his studies to advance our understanding of physiology and medicine, and there are numerous examples of this. A case in point concerns his work on the cardiovascular system. A complete working knowledge of the cardiovascular system is obviously of prime importance for our understanding of medicine and physiology, and so it is worth considering in a little more detail. In this book, I will trace how animal experimentation has been used to investigate the cardiovascular system over the centuries as a touchstone for understanding scientific progress and its dependence on the use of animals.

Aristotle had made some attempts to understand how the heart worked. And he certainly could be very cruel. For example, he invented a procedure where he starved and then strangled animals to try and get a better look at their blood vessels, which were better defined under these conditions. Unfortunately, unbeknownst to Aristotle, when animals are killed by strangulation, the arterioles of their lungs close off, because of asphyxia, while the heart is still beating. As a result, most of the blood is forced into the veins and the right side of the heart. The left side of the heart and the arteries are left largely empty of blood. Because Aristotle didn't know this, his results gave rise to hundreds of years of theorizing about the heart that assumed that arteries didn't contain blood.[8] This is something that will certainly seem strange to a modern reader. How could an entire civilization believe such a thing for so long? How could people be led astray in this way? Isn't the correct answer obvious? So, as we can see, it is all very well to perform experiments, but we need to interpret the results accurately; otherwise they may be misleading rather than helpful. This is still something that applies to animal experimentation today, particularly with things like the development of new drugs where we can easily be led down the wrong path (see Chapter 4).

The first really complete theory as to how the cardiovascular system functioned came from the studies of Herophilus and Erasistratus several centuries prior to Galen.[9] As we have discussed, among other things, these investigators were able to distinguish veins and arteries and show they were different from other structures such as nerves. They suggested that all these long, thin structures acted as channels allowing the distribution of the humors to different parts of the body (Figure 2.1). Erasistratus suggested that food was processed to a liquid by the stomach, from which it escaped through

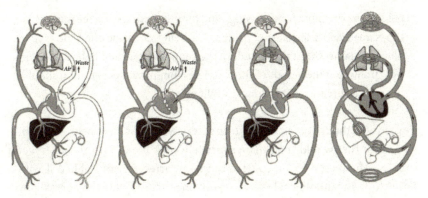

Figure 2.1 Development of theories of the cardiovascular system over time. For details see note 9.

tiny pores traveling to the liver, where it was transformed into blood. This blood then traveled via the veins to all parts of the body, where it produced nourishment and supported the growth and survival of all the different tissues. As was the case with his predecessors such as Aristotle, Erasistratus also believed that the arteries didn't contain blood under normal conditions but rather a spirit called "pneuma," which was something akin to air but also had numerous additional life-giving properties. Most biological scientists in antiquity were "vitalists." They believed that there were unique substances— usually spirits of various types—that inhabited living matter and were not found associated with inert inorganic matter; these spirits were responsible for producing the phenomenon of life. In other words, live animals had to be made of something in addition to elements like carbon, oxygen, hydrogen, and nitrogen; otherwise they would just be dead objects like rocks. Pneuma was a vitalist principle of great importance in antiquity as a life-giving substance, and as we shall see, it greatly informed theories of human and animal physiology.

How, then, did Erasistratus explain why blood spurted out of arteries if they were severed? He wrote as follows:

Arteries and veins, the vessels carrying respectively pneuma and blood, are each successively divided up into a great number of smaller and smaller branches, and are spread to all parts of the body, so that there is no part in which their terminations are not present. They end up in such small vessels, the mouths of which are joined, that the blood contained in the veinlets is held up or prevented from passing. And for this reason, since the mouths or the veinlets and arterioles are continuous, being alongside one another, the blood remains within the bounds of the venous system and in health never penetrates into the vessels of the arterioles. But if such penetration should take place, disease must necessarily follow.

Erasistratus thought that once cut, an artery would lose its pneuma, creating a vacuum that would then inappropriately suck blood into the artery, and that was the reason blood spurted out of an artery if you stabbed it with a knife. It was a pathological response, not one that reflected the normal physiology of the animal. Erasistratus could have performed experiments to test this hypothetical model, which is what a modern scientist would do, but this is something that never seems to have occurred to him.

Two hundred years later, Galen's animal experiments didn't support Erasistratus's cardiovascular model. He did agree that blood was made in the liver, which was deemed to be the source of all the veins, which then supplied the blood to all the "hungry" tissues of the body, where it was used up for nutritive purposes.[10] Hence, to maintain a continuous supply, more blood was constantly required to be made from more food. The model therefore was an open-ended one (see Figure 2.1). Nutrition was important to keep up the blood supply, which would then feed the tissues. A problem here, which we would not have when conducting a scientific investigation today, was that both Erasistratus and Galen failed to address the quantitative aspects of their models; they didn't measure things carefully to see if everything really added up, something that is essential in modern science. How much blood would be needed each day? How much food would be required to make this much blood? Was this a realistic proposition? Nowadays, the actual measurement of substances and the application of mathematics and statistical analysis are absolutely essential to experimentation and establishing scientific "truth." I am free to imagine all kinds of solutions to a scientific problem, but if the numbers don't add up, then I cannot support my hypothesis and need to move on to considering something else.

On the other hand, Galen was unconvinced by Erasistratus's idea that arteries didn't normally contain blood. He therefore conducted what was a real hypothesis-testing experiment, and for this he used animals as experimental subjects. He vivisected animals and ligated arteries in two places so nothing could get in or out. He then cut the intermediary segment and blood spurted out, so he argued the arteries must normally contain blood. Now this was a very innovative kind of experiment and argument, the type of experiment that might well be performed by a physiologist today. With Galen, then, we start to see the emergence not only of the types of experiments that Aristotle might have performed but also more serious attempts to test hypotheses experimentally using animals as functional physiological models for humans.

Galen, then, had experimentally corrected Erasistratus's idea that arteries didn't contain blood. However, in many other respects Galen was wrong when he tried to build a complete model of the cardiovascular system (see Figure 2.1). Galen didn't really think of the heart as a muscle but more as a furnace that produced "innate heat," one of the life-giving spirits that gave the body its warmth and vitality. Some of this warmth was used to help the heart to dilate, and through this action it would draw blood into its right ventricle just as "the dilation of a pair of bellows sucks in air, or as the flames of a lamp suck up oil." Hence, Galen considered that it was the relaxation or diastolic movement of the heart that represented its "active" phase, something that is clearly incorrect according to our modern understanding of cardiovascular function (see Chapter 3). Following its arrival in the right ventricle, some of the blood would flow into the pulmonary artery and nourish the lungs, and a small amount would move through tiny holes in the septum of the heart into the left ventricle, where it could be sucked into the arteries when they dilated. In fact, it was this dilation that Galen thought was responsible for the arterial pulse. Galen made a very detailed study of the pulse and how it changed in relation to cardiovascular disease, one of his greatest and most original achievements. While in the left ventricle, blood would also mix with pneuma, which would add extra life-giving properties to it. Pneuma was transmitted by the pulmonary veins into the left ventricle, where it cooled the innate heat and combined with the venous blood received through the septum. The arteries then delivered blood to structures like the brain, where it was used for functions such as the control of muscle movement. Galen also suggested a method whereby the cardiovascular system could get rid of waste products, certainly an important idea. He thought that these "noxious vapors" were expelled into the pulmonary vein during contraction of the heart or systole and ultimately expelled through the airways.

Of course, this is not a complete description of Galen's cardiovascular model, but it does illustrate how his animal experiments, together with his theoretical insights, allowed him to produce a much more sophisticated explanation of cardiovascular function than anyone had done previously. Some aspects of Galen's model were correct, although in general, as we shall see when we consider the work of William Harvey in the next chapter, most of it was wrong. Nevertheless, Galen's efforts should still be considered "good science," as his detailed model of cardiovascular function made clear predictions that could be tested by further experiments. And that is the way good science often proceeds, from one step to the next, by constantly testing hypotheses and then refining them or coming up with another idea that does a better job of explaining the data. Furthermore, although both Galen and the seminar speaker described in the previous chapter some 1,800 years later were performing experiments on animals, they both hoped that there was a close enough analogy between animals and humans that their animal data would also be relevant to human beings. But the world would have to wait some 1,400 years before further animal experimentation would be used to test Galen's ideas and move the field of cardiovascular physiology forward again. As we shall see, the reason for this was an extended period of time during which scientific research, together with the use of animals for this purpose, went into a precipitous decline.

How should we judge the work of somebody like Galen? Was he a great scientist, a monster, or both? There is no doubt as to the achievements of Aristotle, Galen, Herophilus, Erasistratus, and other scientists of the ancient world. They invented most of what we now easily recognize as a type of scientific method in the way they collected data, framed hypotheses, and sometimes thought of experiments for testing them. This included the way they used animals and how they argued that results obtained from animal experimentation could be extended to humans by analogy. The results of some of their animal experiments made significant contributions to the progress of biological and medical science. Even though many of their conclusions ultimately proved to be incorrect, their scientific methods often looked quintessentially modern, particularly by the time we get to Galen.

But there is also no doubt that somebody like Galen would be seen as excessively cruel as judged by our current ideas of ethics and morality. Did the influence of Aristotle, the Stoics, and others really allow Galen to truly believe that some of the things he was doing were not perceived by his animal victims in ways that we would now believe were terribly cruel? There is a tendency, I think, to forgive people like Galen by saying that things in antiquity were different in terms of the kinds of behavior that were generally acceptable and that his animal experiments would not have been viewed as cruel during the time when he lived. But this really isn't true. There is plenty of evidence that Greeks and Romans often had closely empathic relationships with animals (see Chapter 7). We know this from what they wrote and from how they portrayed humans and animals in their art.[11] They didn't all behave like Galen. So, even in Roman times the type of behavior displayed by Galen, however brilliant he was as a scientist, can only be considered as cruel and inhumane. Galen is really the poster child for researchers in the biomedical sciences today who perform experiments on animals that can provide us with important information but use horribly cruel methods to achieve their ends. And here is the nature of the problem we still face today: does science have to be cruel to be effective?

In conclusion, we can see that by late antiquity animal experimentation was well established as a paradigm for investigation in the biomedical sciences, reaching its zenith of efficiency and cruelty with Galen. We might well assume that the stage was now set for science to advance further in the years immediately following his death around 210 CE—somebody would pick up Galen's mantle and science would progress based on the achievements of the many important investigators who lived in antiquity. But that isn't what happened. Strangely, Galen would be the last great experimental scientist working in Western Europe for around 1,400 years.

Notes

1. Marshall Clagett, *Greek Science in Antiquity* (New York: Collier Books, 2012), 2.
2. Colin Archibald Russell, *The Earth, Humanity and God: The Templeton Lectures Cambridge 1993*, Reproduction in facsimile, Routledge Revivals (Abingdon, NY: Routledge, 2018), 89.
3. Gilbert White and Anne Secord, *The Natural History of Selborne*, new ed. (Oxford: Oxford University Press, 2013), pt. XLIII.
4. Thomas Browne and C. A. Patrides, *The Major Works*, Penguin English Library (Harmondsworth, UK; New York: Penguin Books, 1977), 77–78.
5. H. von Staden, "The Discovery of the Body: Human Dissection and Its Cultural Contexts in Ancient Greece," *Yale Journal of Biology and Medicine* 65, no. 3 (June 1992): 223–241.
6. Margaret L. Antonio et al., "Ancient Rome: A Genetic Crossroads of Europe and the Mediterranean," *Science* 366, no. 6466 (November 2019): 708–714, doi:10.1126/science.aay6826.
7. These references provide an excellent discussion of the life of Galen as well as the experiments that he carried out: Susan P. Mattern, *The Prince of Medicine: Galen in the Roman Empire* (New York: Oxford University Press, 2013); Owsei Temkin,

Galenism: Rise and Decline of a Medical Philosophy, Cornell Publications in the History of Science (Ithaca, NY: Cornell University Press, 1973).

8. Charles Reginald Schiller Harris, *The Heart and the Vascular System in Ancient Greek Medicine, from Alcmaeon to Galen* (Oxford: Clarendon Press, 1973), provides a detailed discussion of theories of the cardiovascular system in Greek and Roman medicine.

9. W. C. Aird, "Discovery of the Cardiovascular System: From Galen to William Harvey: Discovery of the Cardiovascular System," *Journal of Thrombosis and Haemostasis* 9 (July 2011): 118–129, doi:10.1111/j.1538-7836.2011.04312.x.

Development of models of the cardiovascular system over time. Proceding from left to right—Erasistratus and Herophilus discovered that arteries and veins are separate. According to Erasistratus, veins contain blood (grey color), while arteries contain air (white color). Food is taken up in the intestines by the portal veins, delivered to the liver (black color), transformed into blood, and then transported to the vena cava by way of the hepatic vein. From the vena cava, venous blood is delivered to all parts of the body. Some of the blood is diverted to the right ventricle, from where it enters the pulmonary artery to nourish the lungs. Air is taken up in the lungs by the pulmonary veins, transferred to the left ventricle, and distributed to the tissues via the arteries. Fuliginous vapors (waste) are excreted by retrograde flow through the mitral valve and pulmonary vein. Second from left—Galen demonstrated that arteries normally contain blood, not air. Arterial blood is derived from the passage of venous blood through invisible pores in the interventricular septum (shown as interrupted septal wall). Second from right—Colombo (not discussed here) described the pulmonary circuit, in which venous blood in the right ventricle passes through the lungs into the left ventricle and arteries. However, Colombo maintained the ancient Greek view that blood flow in veins is centrifugal (away from the liver and toward all tissues), with only a small amount entering the right heart. Thus, Colombo's system is a hybrid between closed (pulmonary) and open (systemic). Right—Harvey (see Chapter 3) discovered that blood circulates not only in the lung but also around the whole body. An important clue was the presence of valves in the veins (two of them are shown in white). The liver is no longer the source of veins. Rather, the system is driven by the mechanics of the heart (now shown in black). Transfer of blood from arteries to veins in the lung and periphery may occur through direct connections or anastomoses (as shown) or through porosities in the flesh (the latter mechanism being favored by Harvey).

10. Harris, *The Heart and the Vascular System in Ancient Greek Medicine, from Alcmaeon to Galen*; Aird, "Discovery of the Cardiovascular System."

11. Louise Calder, *Cruelty and Sentimentality: Greek Attitudes to Animals, 600–300 BC* (Oxford: Archaeopress: Available from Hadrian Books, 2011).

3

Circular Arguments

This phase of society was brought to a close by the gradual discovery
of the latent powers stored in the all-permeating fluid which they
call Vril. This fluid is capable of being raised and disciplined into the
mightiest agency over all forms of matter, animate or inanimate. It
can destroy like the flash of lightning; yet, differently applied, it can
replenish or invigorate life, heal, and preserve, and cure of disease.
        —*The Coming Race*, Sir Edward Bulwer-Lytton, 1871

Soon after the death of Galen, scientific research in Western Europe came to
a grinding halt for a period of nearly 1,400 years. It began to pick up again
in the seventeenth century, ushering in an era that is generally known as the
Scientific Revolution and has continued at an ever-increasing pace up to the
present day. It is of great interest to note that when experimental biology in
Western Europe began again after its 1,400-year interregnum, it's stepping-
off point was still the work of Galen, even after such an enormous passage
of time. Moreover, the way people thought about the role of animals as research tools was still very much influenced by the ideas of Aristotle, Galen,
and other important scientists of the ancient world. These great experimental
biologists had ultimately not been forgotten, nor had their ideas about the
status of animals. It's just that these ideas had lain fallow for centuries, only
to arise once again filtered through a new set of religious and philosophical
sensibilities. Fourteen hundred years is really an incredibly long time, and
it's certainly reasonable to ask what happened, how the influence of Greek
and Roman scientists was preserved during this period, and why it then
re-emerged to influence the modern world? Part of the answer is that the
cultural context that had originally allowed Galen's work to flourish, a period we call classical antiquity, disappeared only to be replaced by two new
influences—Christianity and Islam.[1] The role of these important religions
in both discouraging scientific research and simultaneously preserving

The Rise and Fall of Animal Experimentation. Richard J. Miller, Oxford University Press. © Richard J. Miller 2023.
DOI: 10.1093/oso/9780197665756.003.0003

some of its key ideas is a complex one. However, among other things, these considerations surely demonstrate that the way we do experimental science, or even if we do it at all, depends on what is going on in the world in general. In other words, although we might think that pursuing science in its current form is inevitable, history tells us that this assumption is unlikely to be correct.

In this chapter I want to discuss several important historical examples of animal-based experimental research taken sequentially from the seventeenth, eighteenth, and nineteenth centuries that were extremely successful in advancing the goals of science and served as an inspiration to twenty-first-century biomedical researchers and the way science is performed today. We will then be able to ask whether experimental methods such as these are still valid or if, as I will argue, they are rapidly being replaced by more humane and effective approaches.

Dark Matters

In the years following Galen's death the Roman Empire was divided into two separate parts. The western portion was based in Rome and the eastern portion in Constantinople. Constantinople was founded by the emperor Constantine (272–337 CE), who was also the first emperor to convert to Christianity, which subsequently became the official religion of the entire Roman Empire during the reign of Theodosius (347–395 CE). The eastern empire thrived until it finally succumbed to the invading Ottoman Turks in 1453 CE, but the power of the western empire rapidly declined, and it formally ceased to exist following the defeat of its final emperor Romulus Augustulus by the "barbarian" Odoacer in 476 CE. The trappings of urbanized civilization decayed following the fall of Rome, and sophisticated learning became the provenance of the church as Western Europe entered the "Dark Ages" for the next 1,000 years. Christian thinking became the major intellectual influence during this time, particularly the teachings of St. Augustine of Hippo.

Born in present-day Algeria, St. Augustine originally embraced Manicheism, a "silk road" faith of Persian origin. However, his mother was a Christian and, following a move to Rome, he converted to Christianity and was baptized by Ambrose, bishop of Milan, in 386 CE. St. Augustine's thinking on the subject of the treatment of animals was completely unsympathetic, something that is of great importance when one considers the enormous influence his opinions were to have on the subsequent development of religion and science. While in Italy, St. Augustine encountered the works of Plato and several Neo-Platonist philosophers and they made a great impression on him, as reflected in his two most important works, *The City of God* and *The Confessions*. Consequently, his view of animals and their treatment depended not only on his interpretation of the New Testament but also the teachings of his ancient Greek forebears. He was particularly influenced by Aristotle's concept of the "great chain of being," in which humans, who, according to Christianity, were made in God's image, were the only truly rational creatures and had a unique place in the world above all the other animals. Indeed, in St. Augustine's mind, God had created the world and all its creatures specifically for humanity to use in any way they thought fit in their efforts to prepare themselves for the afterlife. As far as animals were concerned, he reflected on events described in the New Testament such as the story of the Gadarene swine. In this story, Jesus exorcised a man's demons by transferring them into a group of pigs who then went berserk, ran down a hill, and drowned themselves in a lake. St. Augustine argued that Jesus permitting the Gadarene swine to drown demonstrated that mankind had no obligations to care for animals, writing, "Christ himself shows that to refrain from the killing of animals and the destroying of plants is the height of superstition." Moreover, he also wrote, "there are no common rights between us and the beasts and trees. . . . [W]e can perceive by their cries that animals die in pain, although we make little of this since the beast, lacking a rational soul, is not related to us by a common nature." Clearly, then, St. Augustine was completely in agreement with the idea of human rational and moral superiority over all other animals. Anything he had to say about dissuading people from being cruel to animals was said because he thought it would encourage them to be cruel to each other rather than because of any feelings he had for the animals themselves. Subsequently, other influential Christian thinkers echoed St Augustine's conclusions. In the thirteenth century St. Thomas Aquinas wrote that because animals are intended for the general use of humans, "it is not wrong for man to make use of them, either by killing or in any other way whatsoever." There has been, in fact, until recently, very little in Christian thinking that concerns the compassionate treatment of animals. The fact that animals lacked souls meant that they simply weren't worth considering.

Another effect of St. Augustine's thought was on the actual practice of science. What was the point of trying to understand the nature of phenomena that were just tepid reflections of the true forms of reality? Scientists were just stuck in the cave described in Plato's *Republic*, trying to make sense out of things by performing experiments on their shadows. Surely, there were more important things to do with your limited time on Earth, such as preparing yourself for the hereafter. As St. Augustine was one of the most influential Christian thinkers over the next millennium, his attitudes became the central dogma of the western church, resulting in an intellectual environment that did not encourage the pursuit of science.

Vesalius, Harvey, and the Birth of Modern Biomedical Research

Following the fall of Rome, Galen's works, and many other scientific and philosophical masterpieces of classical antiquity, were "lost" to Western Europe, with a relatively small number being preserved in the monasteries, which became the only real repositories of learning for many centuries. Fortunately, it was also the fate of many of these works to be transported by heretic Nestorian Christians to Persia in the second century CE, where they ultimately became important sources of medical information and practice for the great Arab caliphates that emerged around 700 CE and stretched from Andalusia in the West to the borders of India in the East. It was here, under Arab guardianship, that Galenic medicine flourished for the next 800 years before being "rediscovered" during the Christian reconquest of Spain beginning in the tenth century.[2] To this we can add the major project of the Renaissance in fifteenth-century Italy aimed at discovering lost manuscripts in their original Greek from places such as the Ottoman Empire and translating them into Latin.

A key event was the rediscovery, translation, and publication of *On Anatomical Procedures*, a Galenic treatise wholly unknown to pre-Renaissance scholars but eventually translated in 1531 by Johannes Guinther of Andernach. Really, Galen's book was a primer for teaching the dissection of animals, including a recommendation as to the importance of animal vivisection to anybody who was serious about anatomy and medicine. The book sparked a great deal of new interest in the practice of dissection and anatomy. One of the people who was introduced to *On Anatomical Procedures* was a Flemish student named Andreas Vesalius. Vesalius was an extremely precocious and talented anatomist who, like Galen before him, was passionate about dissecting everything he could lay his hands on whether it was alive or dead. Biographies of Vesalius often relate that as a child one of his hobbies was catching and dissecting cats, not something we normally associate with children. After an early career in Paris and various parts of Europe, he obtained his doctorate at the University of Padua and was quickly promoted to the chair of anatomy at the age of twenty-three. In 1543, when Vesalius was twenty-eight, he published his *De humani corporis fabrica* (*On the Fabric of the Human Body*), usually referred to by its Latin name *De fabrica*. This book was truly a landmark in the history of science. Not only did it go into human anatomy, which had been reintroduced at university medical schools by this time, in incredible and meticulous detail, but it was also aided by the use of magnificent illustrations, which were probably prepared by one of the students of the great artist Titian. The influence of this book on the entire subsequent history of medicine and anatomy cannot be overestimated. By the way, the year 1543 was probably the most important ever in terms of scientific publishing. It saw the appearance of not only *De fabrica* (June 1543) but also *De Revolutionibus Orbium Coelestium* (*On the Revolutions of the Heavenly Spheres*) (May 1543) by Nicolaus Copernicus, the book in which he presented his idea that Earth circled the sun rather than the other way around. A few years later the Italian cleric Giordano Bruno conceived of the universe as being infinite, full of countless stars and planets and lacking a clear center. Clearly, the intellectual climate in Europe was starting to heat up. Old ideas were coming under scrutiny, and the concepts of what we would now identify as modern science would soon be with us. Humanity was about the enter the period known as the Enlightenment, when rational scientific thinking would become the primary influence on the development of human culture for the next two hundred years.

Like virtually everybody else at the time, Vesalius initially had enormous respect for Galen's teachings and didn't set out to correct his conclusions. Rather, Vesalius viewed his work as one of homage and clarification. Nevertheless, by the time his anatomical work had advanced, it was clear to him that there were many problems with Galen's opus. Indeed, it has been estimated that Vesalius identified some 200 or more problems with Galen's work, which he actively pointed out in his books and during his popular public appearances—something that didn't necessarily endear him to the rest of the anatomy department faculty. Among other things, Vesalius criticized Galen's erroneous conclusions as being due to the fact that he never dissected humans—only animals. Clearly, animals didn't always work in exactly the same way as human beings—a consideration we shall return to when we examine the general relevance of animal research to human medicine. Vesalius was one of the individuals who helped to establish Padua as the world's foremost center for medical research in the seventeenth century. He was a worthy successor to Galen, having many things in common with the great Greek doctor in addition to being an outstanding scientist. Both were charismatic characters who craved public attention and were widely known for their extensive and exuberant public performances.

One of Vesalius's major problems with Galen concerned the anatomy and function of the heart.[3] Here, Vesalius was unable to identify the pores in the septum that were supposed to allow the movement of some blood from the right to the left ventricles of the heart as suggested by Galen. Vesalius discussed his observations loudly, publicly, and often, even though many of his colleagues were slow to accept his conclusions. But in the end, it was really Vesalius's critique of Galen that marked the start of the decline of the Greek doctor's influence.

With Vesalius, the edifice of Galenic cardiovascular physiology began to look very shaky, and not before time, considering that our understanding of the topic had not really changed for nearly 1,400 years. And why was that? It seems reasonable to conclude that Galen's model survived because there was no more satisfactory model to replace it: during this entire extended period of time, nobody had performed the types of experimental studies that might have yielded something better.

The man who finally laid Galen's model to rest was William Harvey.[4] Harvey is considered to be a key figure in the history of biomedical research, and the wide-ranging experiments he performed that helped him establish his cardiovascular model were very similar to the way research is still carried out today. He is an outstanding exemplar of the important impact animal research can have on the development of science and medicine.

Harvey was born on April 1, 1578, in the English seaside town of Folkestone and was the son of a wealthy businessman. In 1594, he entered Caius College Cambridge and graduated four years later with a bachelor's degree. The graduates of Caius College were encouraged by its founder, Dr. John Caius, to study abroad, and for medical research in those days, the best place to do this was Padua. On April 25, 1602, Harvey received his diploma of Doctor of Physic from the University of Padua, and his experiences there were to have a great influence on him. He then returned to England, where he became an extremely respected doctor, eventually becoming the private physician to King James I. During this time Harvey was also busy privately conducting a program of experimental research described in his book *De Motu Cordis* published in 1628 (the full title of the book is *Exercitatio Anatomica de Motu Cordis et Sanguinis in Animalibus* [*Anatomical Exercise About the Movement of the Heart and Blood in Animals*]). This seventy-two-page monograph is really a manifesto detailing the use of animals, and vivisection in particular, for carrying out an extensive research program in cardiovascular physiology. Consequently, it is worth looking at some of the things Harvey did if we want to understand how our current use of animals in modern science developed.

While Harvey was in Padua, he attended lectures given by Girolamo Fabrizio (also known as Fabricius, 1537–1619).[5] Fabricius made an important anatomical discovery when he identified structures within veins that we now know act as valves to help control the direction of blood flow through these vessels, something that would ultimately have an important influence on Harvey's thinking. Fabricius published a description of these structures in 1603, although at the time, he didn't really know what they did. As described in his account, he referred to them as *De venarum ostiolis* (the little doors of the veins). Fabricius did perform some rudimentary experiments to try and understand what these "little doors" did. He ligated the arms of human subjects using a tourniquet and rubbed the veins that popped up to see what happened. He found that manipulations of this type failed to increase the movement of blood from the arm toward the hand in any obvious way. Hence, as an adherent of Galenism, Fabricius proposed that the function of the "little doors" was to slow the centrifugal flow of blood through the veins toward the periphery—that is, toward the hands in these experiments. In other words, they served to prevent excessive outward movement of blood, which might be wasteful and cause an imbalance of the blood supply to different tissues. This was only slightly different from Galen's view, which posited that the flow of blood was controlled by the power of the "hungry" tissues to attract blood according to their needs. In other words, Fabricius didn't understand that these structures were really valves and that they controlled the actual direction of blood flow rather than just the volume and rate of blood flow. But one of Harvey's great insights, as we shall see, was that he did come to understand the manner in which these valves functioned as part of an overall integrated theory of how the entire cardiovascular system worked.

Harvey's ultimate model of the cardiovascular system depended on his precise use of experimentation to test his hypotheses so that one conclusion built upon another. Animal dissection together with some imagination and reasoning may suggest how things function in practice. However, what

Harvey's scientific method added to the mixture was the actual testing of an idea through the use of some artificial, that is, experimental, intervention with a predicted outcome to see if an idea was really true or not in practice. Galen had done some of this, of course, but not as extensively as Harvey and not iteratively to test all facets of his model as it developed. When Harvey published *De Motu Cordis* he said, "I profess to learn and teach anatomy not from books but from dissections, not from the tenets of philosophers, but from the fabric of nature." Clearly, he considered anatomy as his starting point, but he was determined to go further in his search for the truth.

Sometime after 1610, in his house in Ludgate in London, Harvey set up his own personal scientific research institute. Of course, Harvey was independently wealthy and had the means to do this. For example, in addition to laboratories, he housed his own menagerie for vivisection purposes. The following animals were employed or at least mentioned by Harvey in connection with his research program: toads, crabs, shrimp, whelks, oysters, fish, lizards, tortoises, serpents, fowls, pigeons, geese, rabbits, mice, sheep, pigs, dogs, dogfish, sharks, tadpoles, crocodiles, ostriches, pheasants, wolves, horses, goats, and elephants.[6] Indeed, like Vesalius before him, Harvey was keen to dissect anything he could lay his hands on, including, in addition to the above list, the corpse of his dead father and his wife's pet parrot. Using this huge variety of animals, as well as human subjects, Harvey carried out a series of experiments between 1610 and the early 1620s aimed at testing different aspects of Galen's model of the cardiovascular system.[7] Every part of Galen's model suggested particular consequences, and Harvey would design experiments to see if these were true or not. The main experimental interventions he used were cutting and tying. If a certain vessel was ligated, where did the blood build up? What happened if you cut a particular vein or artery? What did Galen predict would happen and what happened in actuality? These artificial interventions are the essence of experimentation— you intervene in a system in some way and see if you can predict the result. For example, Harvey had a man who had just been put to death by hanging brought immediately to his laboratory. Having laid the fresh corpse out on his experimental bench, Harvey cut into his thorax, exposing his heart. He then ligated the pulmonary artery (Figure 3.1). Harvey inserted a tube into the vena cava, the large vein leading into the right side of the heart, and forcefully injected water into it. If there were pores in the septum of the heart, as dictated by Galen's model, then this procedure should have caused some water to appear in the left ventricle.

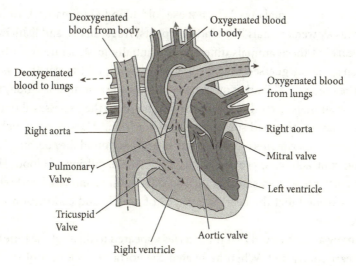

Deoxygenated
blood from body

Oxygenated blood
to body

Deoxygenated
blood to lungs

Oxygenated blood
from lungs

Right aorta

Right aorta

Mitral valve

Pulmonary
Valve

Left ventricle

Tricuspid
Valve

Aortic valve

Right ventricle

Figure 3.1 Depiction of the direction of blood flow around the human heart.
Image from Shutterstock.

Harvey observed that the right ventricle swelled, but "not even a single drop escaped through the septum into the left ventricle—by my troth there are no pores, nor can they be demonstrated," wrote Harvey. Next, Harvey released the ligature from the pulmonary artery and tried again. Now, water almost immediately "shot forward, mixed with a large amount of blood out of the pulmonary artery and into the left ventricle." Hence, blood must be able to pass from the right ventricle of the heart into the lungs and from there into the left ventricle.[8]

Harvey also vivisected dogs and ligated the vena cava. When he did this, the pulmonary artery, pulmonary vein, and aorta emptied of blood and collapsed, but the vena cava swelled up as if it was trying to force blood through the rest of the heart. When he removed the ligation, the heart, lungs, and all the vessels filled with blood once more, again supporting the view that blood could move from the right ventricle to the left via the lungs.

What about the movement of the heart itself? Which was the active phase of the heart, contraction (systole) or relaxation (diastole)? This was something that was difficult to test directly in humans, so Harvey vivisected animals, literally thousands of dogs, cats, rodents, and other mammals—and don't forget this was completely without any kind of anesthetic. However, the heartbeat he observed in these creatures was generally so rapid that it proved difficult for him to come to a definitive conclusion. Then Harvey had a brilliant experimental idea. Why not use cold-blooded animals, whose heart-beat was slower, animals like toads, serpents, frogs, snails, and fish? Even if the hearts of these animals differed substantially in structure from that of humans, Harvey guessed that at some very basic level the way they worked must be the same. Scientists often make these same assumptions today when they use animals as models for humans. Sometimes their guesses about their similarities are true and sometimes they aren't, but Harvey got it right. He conducted experiments on eel and fish hearts and found they became harder during contraction, appearing white due to the expulsion of blood. He cut hearts into pieces and saw that they spontaneously contracted and relaxed, so he reasoned that the heart was a type of muscle and contraction was its active phase.

Harvey also observed that the arteries appeared to dilate at the time when the heart contracted. When he ligated the aorta, the left part of the heart swelled enormously, and the aortic pulse ceased. Then, when he released the ligation, the pulse resumed. Harvey concluded that the pulse was due to the "impulsion of blood" on contraction of the heart. When Harvey severed the arteries of dogs, he saw that blood spurted out from the severed vessel at the same time that the heart contracted. Thus, in contrast to what Galen believed, contraction (systole) of the heart occurred simultaneously with dilation of the arteries. The arteries, therefore, are normally distended because they are filled like sacs or bladders, as opposed to expanding like bellows, as Galen had thought.

Harvey also repeated and extended Fabricius's experiments, ligating the arms of his servants with light tourniquets, and like Fabricius, he observed that he couldn't force blood to move toward the hand. In addition, he performed experiments in which he ligated the exposed veins of vivisected animals a little above a pair of "little doors" and saw that in spite of a buildup of blood between the doors and the ligature, no degree of force would cause the blood to flow backward. Harvey therefore came to an important conclusion that differed from that of Fabricius, which was that the little doors were valves that controlled the direction of blood flow so that blood in the veins always flowed toward the heart and never away from it toward the periphery. Indeed, it was the valves in the veins or in the chambers of the heart that directed the flow of blood, and Harvey's understanding of this function was really an essential aspect of his model of the cardiovascular system and something of which Galen had been completely unaware.

There was one other key piece of information that Harvey considered. He asked a quantitative question, something else that is key to good experimentation and which nobody had previously asked: "How much blood does the liver need to make to provide all of the blood that is used by the body every day?" Harvey estimated that each time the heart beats, it uses two ounces of blood during systole. Because the heart beats on average seventy-two times per minute, he calculated that 8,640 ounces (or 540 pounds) of blood should be pumped per hour, which was four times the weight of an average human being. It seemed extremely unlikely that the liver could make that much blood. What is more, Harvey showed that the properties of the blood taken from the veins and arteries were basically the same. When removed from the body and allowed to sit in a basin, both kinds of blood had the same color and consistency and coagulated in the same way. Harvey did admit, however, that although arteries and veins contain the same blood, the arterial blood was more "spirituous" and possessed a greater life-giving vital potential. He was not to know at this point in time that this extra secret ingredient was oxygen and, in fact, that the main purpose of blood and the pumping of the heart was to enable the delivery of oxygen to the "hungry tissues" and the removal of spent oxygen in the form of carbon dioxide back to the lungs, from which it could be exhaled.

In conclusion, the use of an integrated and extensive program of human and animal experimentation allowed Harvey to arrive at a model in which the blood was reused by circulating around the body—from the right ventricle through the lungs and then to the left side of the heart. Contraction would force the contents of the left side of the heart through the arteries to the tissues. Once the blood was spent, it would drain back to the right side of the heart through the veins. This model is essentially correct, and since Harvey's time, only details have been added to it. Nevertheless, many of Harvey's contemporaries were still very invested in Galen's ideas, and they weren't going to give them up easily. Harvey's detractors attacked him. How did the blood that returned to the heart move from the arteries into the veins? There had to be some connection between these vessels, but nobody had ever seen such things. But the connections were there; small vessels called capillaries were discovered a few years later in 1661 by Marcello Malpighi with the aid of a fabulous new invention—the microscope. Perhaps the violent, painful deaths suffered by research animals interfered with natural conditions? Some asked whether Harvey's findings in animals really could be extrapolated to humans? Questions like these are still discussed by scientists today when they wonder about the appropriateness of results from animal experiments and their application to humans.

How are we to judge Harvey? It cannot be denied that he was one of the greatest experimental physiologists who ever lived—many would argue the greatest. Harvey's elucidation of the correct mechanism by which the blood circulates is one of the most important concepts of physiology, and it has been essential for the development of medicines and surgeries for treating human (and animal) cardiovascular disease. On the other hand, Harvey, like Galen before him, was terribly cruel. Harvey inherited his attitudes about animals from the time of ancient Greece, when it was concluded that animals were not morally equal to humans and that they could be used purely as resources. Harvey acted with no apparent concern for the fate of the animals he vivisected, many of which must have died in paroxysms of pain. Science is often just as cruel today, and many experimental programs carried out in contemporary laboratories are extremely similar to Harvey's in many important respects. He was indeed thoroughly modern in his approach, far ahead of his time, which is one of the reasons he was such a great scientist and so influential. It cannot be denied that Harvey's experiments resulted in extraordinarily important medical advances that benefited the human race. Just like Harvey, scientists today justify what they do by saying that they are discovering important things that will help humanity. Animal experimentation is just the price we must pay. We trade our suffering for the suffering of animals, creatures we deem to be of lesser importance. Although some things have changed about the way we use animals experimentally since Harvey's time, we should ask whether in this day and age such an equation is still inevitable or whether there are now other ways of performing experiments and accumulating knowledge. Harvey had few alternatives to the use of animals, but, importantly, that is something that is no longer true. Moreover, is it true that all the work done by scientists today is beneficial to humanity, and if so, how much of it is? Finally, and perhaps most importantly, even if experiments like Harvey's produce important results, are they ethically reasonable?

Descartes and the Animal Machine

Harvey's work clearly established vivisection as an essential experimental paradigm for modern biomedical research. However, Harvey still had one foot in the old world, including its vitalist concepts. Biomedical research needed new philosophical approaches to go along with Harvey's revolutionary experimentation. And these were soon to be supplied by his younger French contemporary Rene Descartes. Descartes raised the question: "Are animals a sort of machine?" And, if that is the case, "What are the implications of such an idea?" When we consider animal experimentation, this is important because, if an animal is just a machine devoid of the capacity for reason, thoughts, and feelings, we might not have to worry so much about what we did to it and could consider its apparent display of emotional responses as just the result of a sophisticated arrangement of interlocking clockwork mechanisms, like a dancing figure in a music box. Ideas like these became extremely influential in biological research over the years following Descartes's death and still are today. Apart from the ethical issues raised by these considerations, they also have important scientific consequences. If animal machines are fundamentally different from human machines, that is, if something essential is missing from them, then how do these differences influence the manner in which we use animals as "models" for representing humans in science, and how do we "translate" results from animal studies to humans?

René Descartes was born in La Haye-en-Touraine, France, on March 31, 1596. Owing to the death of his mother soon after his birth, he was mostly brought up by relatives. Descartes lived a very peripatetic life, joining the army and residing in many different parts of Europe. While on his travels he probably visited many of the continent's palaces and gardens. These were the great showplaces of the time, with aristocrats trying to outdo one another by transforming their gardens into fabulous theme parks. Many of them exhibited incredible automata designed by brilliant hydraulic engineers.[9] Driven by water engines, they displayed famous scenes such as incidents from classical mythology. What impressed Descartes the most was the fact that these automata were not just like clockwork fountains that would spray water on cue at particular times of the day, but were responsive to specific inputs, such as the presence of a human visitor who would trigger an action, suggesting sentience and responsiveness. Surely, thought Descartes, the human body must work in exactly the same manner—and I really do mean in exactly the same manner. He wrote, "Indeed, one may compare the nerves of the machine I am describing with the pipes in the works of these fountains, its muscles and tendons with the various devices and springs which serve to set them in motion, its animal spirits with the water that drives them, the heart with the source of the water, and the cavaties of the brain with storage tanks." The body, then, was just a machine that worked through a set of reflexes. The behavioral output of the machine could range from something as simple as the withdrawal of a limb to something much more complex such as an emotion or at least the physical actions associated with it. Descartes didn't wish to imply that the human machine was inanimate—far from it. The human body was a machine that was also alive.[10] Understanding the precise mechanisms that contributed to it being alive was an interesting question. And, as Descartes realized, humanity now had experimental approaches, pioneered by people like Harvey, for properly investigating this kind of question. This scientific discipline, for which Harvey's studies were powerful examples, now had a name given to it by the French scientist Jean Fernel. It was called physiology—the understanding of the mechanisms that enabled living systems to function. Descartes thought that all aspects of the way the human body functioned were accessible through this kind of analysis. It was just that the human machine was much more complicated than anything in the gardens of the Hortus Palatinus or Saint-Germaine-en-Laye, because it had a much better designer—almighty God. Descartes realized, as Aristotle had, that living machines had something missing from them, and that was the capability to think rationally so that the machine could receive directions that resulted from thinking and self-reflection, something that was initiated from within. It was the *cogito ergo sum* of the human being. Descartes envisaged this as the rational human soul, something that was completely disembodied, immaterial and distinct from the human machine but that could interact with it and direct it. Indeed, Descartes thought this interaction occurred at a specific point below the brain—the pineal gland. As opposed to humans, animals lacked this God-given self-reflective "soul."

Descartes's influence spread throughout Europe. There were many people who responded to Descartes's work by concluding that animals really were *betes machines* that didn't feel anything at all—even in the face of everyday evidence to the contrary. There are stories such as the one about the philosopher and follower of Descartes, Nicolas Malebranche, who was seen kicking a pregnant dog and saying, "Who cares; she can't feel anything anyway." Or as another follower of Descartes declared, "Animals eat without pleasure, cry without pain, grow without knowing it, desire nothing, fear nothing, know nothing." Such attitudes became the way scientists began to conceptualize animals and their proposed role as experimental models that reflected the nonsentient mechanical aspects of human life.

Deconstructing the Animal Machine

In England, William Harvey became the personal physician to King Charles I and followed the king when he settled in Oxford in 1642 during the civil war. Oxford at the time became a mecca for some of the greatest English scientists. Among the group were Robert Boyle[11] and Robert Hooke,[12] without doubt two of the greatest and most influential scientists in history. Boyle was a wealthy aristocrat, the son of the Earl of Cork, whereas Hooke was the son of a clergyman from the Isle of Wight. Boyle was an extremely devout Christian, something that greatly influenced his work, and he was always at pains to demonstrate how the wonders of Nature reflected the beneficence of the Almighty.

Boyle was interested in research for its own sake in addition to whatever it could usefully accomplish in a directly practical sense. In order to "satisfie mens Curiosite of understanding,"[13] animal experimentation was one of the most prominent methods available to him, and he used it extensively, something that was true of many of his contemporaries. Many of these studies involved vivisection, which was as cruel as that practiced by Harvey, although Boyle appears to have been of two minds about how his victims perceived his actions. Nevertheless, he concluded that whatever he was doing was for the greater glory of God and the human race and so was worth the price—a thoroughly modern attitude.

Whatever his reasons, there can be little doubt as to Boyle's effectiveness as a scientist and his achievements in the fields of biology, chemistry, and physics. Boyle's writings were extremely extensive and detailed, and there are many clearly noted examples of his scientific procedures. One of the impressive things about Boyle's experimental program was his development of many novel pieces of equipment. Here he was often assisted by Robert Hooke. Hooke was one of the most talented individuals in the history of science. He not only was a brilliant theoretician (he told everybody that he discovered gravity before Newton) but also had fantastic ability as an experimentalist and was one of the greatest scientific illustrators. Hooke was a student at Oxford, and after finishing his degree, he was employed by Boyle as his assistant. Boyle quickly became aware of Hooke's talents, and they maintained a lifelong collaboration. Together Boyle and Hooke made a formidable team. Boyle relied on Hooke to come up with equipment for use in his experiments. Probably the most famous example of this was the vacuum pump, or *Pneumatik Engine,* that Hooke built for showing how different physiological processes depend on the availability of air. The equipment consisted of a glass chamber in which an animal or other object could be placed while a pump evacuated the air from it. Not surprisingly, the animals, which could be mammals, birds, reptiles, or insects, were slowly asphyxiated. Not only was this demonstration generally copied and so became widely known, but also it was eventually the subject of a great painting—*An Experiment on a Bird in the Air Pump*—painted in 1768 by Joseph Wright of Derby at a key point in history when the Scientific Revolution was well under way and Britain was on the cusp of the Industrial Revolution (Figure 3.2).

By this time interest in science, including general participation in experimental work, had begun to filter down from the wealthy aristocracy to the burgeoning middle classes. In the scene shown in Figure 3.2, the air pump is being used to suffocate a cockatiel as part of a family's evening of scientific self-instruction. Clearly experiments like those depicted in the painting provided an evening's entertainment for many middle-class families in

Figure 3.2 *An Experiment on a Bird in the Air Pump*—painted in 1768 by Joseph Wright of Derby.
Image from Alamy.

eighteenth-century England. The fact that the purposeful killing of an animal seemed to have made little difference to the men present is something that we can attribute to the influence of thinkers like Descartes. On the other hand, the two girls who are visibly upset by the proceedings represent a different attitude, the idea that the experiment was simply cruel. Soon this new view of science would find an outlet in the antivivisection movement that gained force in Britain in the next century (see Chapter 7).

In time, the mechanical model would develop beyond Descartes's original concepts, and this would have important implications for the use of animals as experimental subjects. Why not consider that the entire human being, including the capacity for rational thought and self-consciousness, was the result of mechanical processes? This viewpoint was clearly committed to paper by Julien Offray de La Mettrie in his book *L'homme machine (Man the Machine*, 1747),[14] where he provided a completely materialist view of humanity. A human being was a mechanical animal but instead of being made of clockwork, it was made of organic matter that organized and energized itself. There was nothing special about a human being; there was no divinity about it. There were quantitative not qualitative differences between humans and animals, implying that all aspects of humanity were potentially susceptible to scientific experimentation and explanation. Hence, one might also imagine that all the properties of a human could be reflected in animal models, or, put another way, experimental animal models might be useful analogies for studying all aspects of human biology. As La Mettrie himself said, "Man in his first principle is nothing but a Worm." Or as Charles Darwin's grandfather Erasmus, a core member of the Lunar Society, one of the Enlightenment's most important scientific discussion groups, wrote, "Go, proud reasoner and call the worm thy sister!" These are prophetic words considering that in contemporary biomedical research, one of the most commonly employed animal analogs or models of a human is the nematode worm *Caenorhabditis elegans* (see Chapter 4).

From the point of view of the development of modern biomedical research, a key conceptual and technical advance was made by the Swiss physiologist Albrecht Haller (1708–1777), who performed experiments in which he vivisected animals (several thousand of them of varying types) and stimulated different organs with chemicals, with heat, or mechanically. If a stimulus evoked pain, he deemed the tissue "sensible," and if it elicited muscle contraction, he deemed it "irritable." He concluded from these studies that only nerves had the property of sensibility and that muscles had the property of irritability. Importantly, he observed that the property of irritability was still manifest even if the innervation of a muscle was cut because muscles could still be made to contract by stimulating them directly. Hence, muscle irritability implied that muscles had an internal propensity to contract if they received the appropriate stimulus. The way in which Haller conducted some of his experiments was novel because he observed that the property of muscle irritability remained intact even when the muscles were completely removed from the body of an animal and subsequently stimulated in isolation. This was extremely important because it implied that there was the possibility of studying the properties of animal tissues once they had been removed from their normal environments within the intact animal. This is what we now refer to as an "isolated tissue preparation," in which a dissected piece of tissue can be removed, studied, and manipulated separately from the original animal. It was a piece of Descartes's animal machine that could be studied in isolation, just like a part of a watch or bicycle. Here, then, was the first step in taking the animal machine apart and examining its different components, really the beginning of the modern sciences of experimental physiology and pharmacology. The isolated tissue preparations that have been used ever since and have been absolutely essential to the progress of things like drug development are the descendants of Haller's work. Such procedures are still widely used today. Sir John Vane, for example, who won the Noble Prize in 1982 for elucidating the mechanism of action of aspirin-like drugs, was famous for his experiments that used "cascades" of isolated tissue preparations taken from the organs of different animals to detect the presence of an important physiological mediator in a biological fluid.

One type of animal that turned out to be particularly useful in this regard was the frog. Frogs were just complicated enough to exhibit behaviors like negative reactions to painful stimuli but were simple enough to be easily dissectible, their powerful leg muscles being ideal for the study of muscle contraction. Importantly, contractions of isolated frog muscles were observed to last for a considerable time following an animal's death. This is extremely useful from a practical point of view because the longevity of a frog preparation allowed extended studies to be carried out consisting of multiple replications of experiments over a long period of time. By the middle of the eighteenth century, frogs had become so popular as experimental animals that some wit quipped that it reminded him of the massacre of the frogs in the pseudo-Homeric *Batrachomyomachia*.[14]

The Galvani/Volta Debate

One area of scientific investigation that was of great interest in the eighteenth century concerned describing the biological properties of electricity, something that had been pioneered by investigators like Benjamin Franklin. Indeed, electrified humans and animals performing all manner of interesting tricks were very popular as demonstrations for the public at that time and played to huge audiences.[15] A major advance in the field had been the development of early electrical capacitors, such as the Franklin square or Leyden jar, an apparatus for storing electricity, allowing the "electrical fluid" to be transported from place to place and to be discharged on demand. Following Haller's investigations, it became clear that stimulation of muscles with an electric charge was one of the most effective ways of engaging their irritability and inducing contractions. Two eighteenth-century Italian scientists applied electrical stimulation to frogs, observing that "all the parts of frogs, either live or dead, do contract with electric sparks" and that the electrical stimulation of the nerves that ran from the spinal cord to the leg muscles produced contractions of the inferior limbs even when other mechanical or chemical stimuli were ineffective.[16] In some experiments they dissected frogs and left only their inferior limbs attached to their spinal nerves; in other words, they produced an "isolated nerve-muscle preparation," consisting of the spinal cord and legs (Figure 3.3).

They then demonstrated that the legs would twitch (contract) when they stimulated the preparation electrically. Indeed, this isolated preparation proved to be extraordinarily sensitive to electrical stimulation. But what was the true physiological significance of such an observation? Was this just another party trick for entertaining the public, or did electricity really have something to do with the properties of the "nervous fluids" that normally helped to control muscle contraction in a live frog? Interestingly, at around this time, investigators had shown that certain fish such as the Mediterranean ray *Torpedo torpedo* or the South American eel *Electrophorous electricus* were capable of making and storing their own electricity, which they could discharge as an electric shock, just like a Leyden jar, to kill prey or discourage predators.[17] You could even get an electric fish to discharge its stored electricity and produce a visible spark—clearly evidence of real electricity to an eighteenth-century investigator. However, most scientists thought that this was just an isolated phenomenon and that other types of animals couldn't produce their own electricity. This was to be the subject of the great debate

Galvani frog experiment

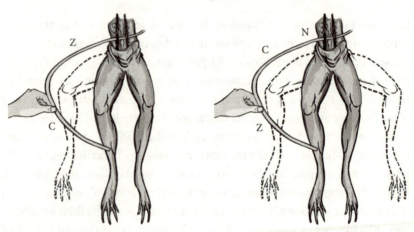

Figure 3.3 Isolated spinal cord and frog leg preparation as used by Galvani and Volta in their experiments. C, copper. Z, zinc. N, nerve.
Image from Shutterstock.

between the Italian scientists Luigi Galvani and Alessandro Volta,[18] in which experimental animals would play a central role.

Around 1780 Luigi Galvani initiated a series of experiments in his laboratory in Bologna that would result in his discovery of "animal electricity." Galvani's initial results were described in his publication *De viribus electricitatis in motu musculari (On the Forces of Electricity in Muscular Motion)* in 1792, which, like Harvey's *De Motu Cordis*, is one of the key publications in the history of biomedical research. To begin his research program, Galvani repeated observations that had already been made by his colleagues showing that electrical stimulation of the muscles or nerves could produce a frog leg twitch (contraction) response. He would deliver an electrical stimulus using a Franklin square or Leyden jar connected by a wire to the part of the preparation he was testing. One thing that did surprise him was the sensitivity of the preparation, particularly when a stimulus was applied to the nerves. He could use a Leyden jar that was practically depleted of electrical charge and it would still make the preparation twitch effectively. He was therefore able to conclude that, whatever the role of electricity was in normal nerve and muscle physiology, his frog preparation was a fantastically sensitive detector of electricity and, in fact, could be used for that purpose.

Indeed, the preparation was subsequently used as an assay for electricity right up until the nineteenth century when sensitive galvanometers were developed for this purpose.

In January 1781, Galvani made an observation that would be of the greatest significance. While experimenting in his laboratory, his wife, who was also his assistant, touched a Franklin square with her finger, eliciting a spark discharge. At exactly the same moment, Galvani happened to be applying a metal lancet to the nerves of a frog preparation situated at some distance across the room and observed a clear twitch response even though there was no wire connecting the two. Rather than assuming this was just a meaningless result, Galvani was intrigued and followed it up, repeating it several times. As he wrote:

> When my wife or someone else brought a finger close to the conductor [of the machine] and elicited sparks [from it], and at the same time I rubbed the crural nerves or spinal cord with an anatomical knife provided with a bone handle—or even if [the knife] was just brought close to the said spinal cord or nerves—then there were contractions even though no conductor was applied to the glass where the frog lay.

Needless to say, Galvani was puzzled by this result. It was completely at odds with his normal conclusion that electricity flowed from its source through a wire connection into the frog preparation and elicited a contraction. So, where was the electricity coming from, or was something else going on? Was it perhaps that the electricity didn't need to come through a wire connection and that, like Haller's phenomenon of irritability, it was already contained within the frog itself? Indeed, could electricity explain irritability?

Galvani was subsequently distracted from his studies for a time owing to administrative and other tasks but eventually got back to them in 1786 and took his experiments outside. At the time there were various opinions about the true nature of electricity including the possibility that "natural" electricity, as produced in thunderstorms, differed in some way from electricity produced artificially by a machine in the laboratory. Galvani connected his frog preparation to a lightning conductor during stormy conditions and observed that every time there was thunder, the frog legs contracted. Perhaps this wasn't too surprising, but what happened next was. As Galvani described it:

In early September, at twilight, we placed therefore the frogs prepared in the usual manner horizontally over the [iron] railing. Their spinal cords were pierced by iron hooks, which were used to suspend them. The hooks touched the iron bar. And, lo and behold, the frogs began to display spontaneous, irregular, and frequent movements. If the hook was pressed against the iron surface with a finger, the frogs, if at rest, became excited—as often as the hook was pressed in the manner described.

Galvani then took his experiment inside again and tried just connecting the nerves and leg muscles with pieces of metal. Once again, as soon as he completed the circuit, he observed a twitch response indicating that it had nothing to do with absorbing electricity from the atmosphere. After thinking about the various possibilities, Galvani concluded, "This outcome surprised us greatly and at the same time led us to the suspicion that there was a kind of electricity inherent in the animal itself." Here, then, was more evidence for the possible existence of "animal electricity" but now in a frog, an animal that was very different from an electric fish or eel. Further investigations indicated that contractions could be elicited by using a metal arc composed of two different metals that connected two different parts of the muscle, two different parts of the nerve, or the nerve and the muscle, this final configuration being the most effective.

Like Harvey before him, Galvani came up with a mechanical analogy to help him visualize what was going on. For Harvey, the heart was like a pump; for Galvani, the frog preparation was like a Leyden jar, which produced a twitch when it was discharged (Figure 3.4). As Galvani himself wrote: "There is no doubt that one sort of electricity is located in the muscle, the other in the nerve; in order to put each of them into action it is sufficient to apply an appropriate conducting body to the nerve and the muscle so as to establish a good communication between them."

If the Leyden jar analogy was correct, then it implied that there must be an electrical disequilibrium (what we nowadays would call a potential difference) present within the frog tissues. According to his model, this electric disequilibrium was located between the internal and the external surfaces of the muscle fibers, with the nerve fibers acting like the metal rod that draws electricity into the Leyden jar. When a metal arc was applied to the frog preparation, electricity would be free to run from the interior to the exterior of both the Leyden jar and its muscle equivalent, producing an electric current (Figure 3.4), which, as we have discussed, was a powerful stimulus of

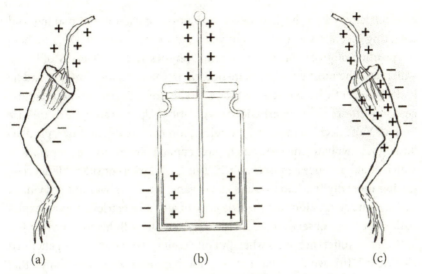

Figure 3.4 Theoretical comparison of the disposition of electrical charges between a Leyden jar (b) and Galvani's frog preparation.[17]
Image courtesy of Oxford University Press.

muscle contractions. One question was, how could this disequilibrium be maintained given the normally aqueous environment of the muscle that would be expected to discharge it? Here again, Galvani had a wonderfully modern idea. He suggested that the muscle was covered with an "oily" non-conducting layer that would serve to keep the disequilibrium in place, and as we now know, this is the role of the cell membrane, which has precisely this nonconducting quality.

One difference between Harvey's situation and Galvani's was that the latter met with considerable scientific opposition to his work. Of course, many people objected to Harvey's conclusions as well, but this was more in the way of differences in interpretation rather than the presentation of experimental data obtained by a rival investigator. In Galvani's time, more "gentlemen" were performing true scientific experiments in their own laboratories and could be tempted to test Galvani's conclusions for themselves. And that is precisely what happened. The publication of Galvani's results started a lively debate among a large number of scientists from all over Europe who became engaged in efforts to replicate them.

Among these was his compatriot Alessandro Volta. Volta was already a very famous scientist, particularly noted for his work on the properties of electricity, where he had developed novel apparatus for making and detecting it. Volta was forty-seven years old in the spring of 1792; he held the position of professor of experimental physics at the University of Pavia, which was the most prestigious institution in the Habsburg Empire, and he had also recently been elected a Fellow of the Royal Society, something that certainly attests to his international reputation. Upon reading Galvani's *De viribus*, Volta decided to attempt to repeat some of the key findings, particularly those involving the use of metal arcs applied directly to the preparation, which could produce a muscle twitch. This he did with considerable success, getting precisely the same results as Galvani. But Volta was not convinced by Galvani's suggestion that the existence of animal electricity was an explanation for these observations. Rather, he came up with his own novel idea. What if two different metals when put into contact with each other generated electricity? This wouldn't have to be a lot, because it had been shown that the frog preparation could act as an ultra-sensitive electricity detector. Then, supposed Volta, there would be no reason to postulate the existence of animal electricity, just a small amount of "metal electricity" that would trigger the frog preparation to twitch. To support this idea Volta did something interesting. He knew that when an electrical discharge was applied to his tongue, it produced an acid-like taste. So, instead, he applied two different metals to his tongue, just as he would have applied to the frog preparation, and got exactly the same acid-tasting response, clearly indicating that the two metals could act together to produce electricity. Volta then proceeded to carry out a series of experiments in which he tested different combinations of metals to see which ones were the best at triggering a frog leg twitch. Volta was naturally keen to establish the primacy of his new discovery, saying that he had discovered "a new virtue of metals, never supposed yet, which my experiments have led me to discover."

How did Galvani propose to counter Volta's arguments against the existence of animal electricity and the primacy of metal electricity? He performed experiments in which he could dispense with using the metal arc altogether. Instead, he took the end of the spinal nerve and carefully manipulated it so that it curved around and touched the leg muscle. As soon as contact was made, a contraction was observed (Figure 3.5).

Volta shot back. He supposed that it didn't necessarily have to be two different metals that generated electricity, but that the coming together of any two different conducting materials could do the same thing. Of course, the nerve endings and the leg muscle were two different conducting materials,

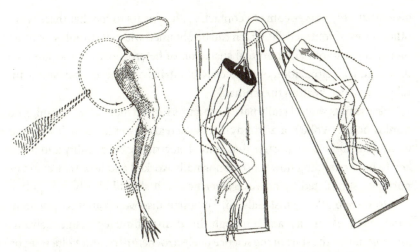

Figure 3.5 Examples of initiating a leg contraction using Galvani's frog preparation without a metal intermediate.[17]
Courtesy of Oxford University Press.

so just like an arc of zinc and copper, they too would produce electricity and that would stimulate the frog preparation.

Over to Galvani: he thought up a very crafty new experiment:

> I prepared the animal in the usual way; then I cut the one and the other of the sciatic nerves near their exit from the vertebral canal; afterwards, divided, and separated a leg from the other, in such a way that any of them would remain with its corresponding nerve; in the following I bent the nerve of one in the shape of a small bow, and after that having lifted with the usual small glass rod the nerve of the other leg, I let it fall on that nervous bow. In doing that one should observe the precaution that, in its fall, the nerve should touch in two points the other nerve bent as a bow; and, moreover, that the small mouth of the first nerve is one of the two points. I saw then the moving of the leg corresponding to the nerve that I let fall onto the nerve of the other; sometimes even both legs. The experiment succeeds while the two preparations are totally isolated one from the other, and have no reciprocal relation, except for the touching of the nerve.

As he described it, Galvani had succeeded in obtaining contractions by forming an arc made exclusively of nerve substance. He concluded: "What heterogeneity could then be invoked in support of the occurred contractions, given that only nerves come to contact. . . . It appears to me that there is another series of contractions which can be obtained without stimulus, without metal, and without any minimal suspicion of heterogeneity; these are then produced by a circulation of an electricity intrinsic to the animal, and naturally unbalanced within the animal."

These results should really have been game set and match for Galvani. And frankly neither Volta nor anybody else had a real answer to them. It was clear to Galvani and his supporters that animal electricity was a reality and it was key to understanding how muscles normally worked and how they could be normally stimulated by nerves. Nevertheless, it should also be recognized that to wage battle against Galvani, Volta had discovered another phenomenon, metal electricity, and although this didn't ultimately refute Galvani's findings, it was the start of the science of electrochemistry, and Volta went on to capitalize on his findings. He devised a piece of equipment that involved stacking pairs of different metal rings separated by moist materials on top of one another so that they generated a constant and easily variable source of electricity—it was a "voltaic pile," the first battery, and one of the greatest discoveries in the history of science. (It was actually inspired by the electric organ in torpedo fish in which cellular elements could be observed piled on top of one another.)

The Galvani/Volta debate is a wonderful example of how science makes progress through a constant struggle between hypotheses competing with one another through the use of experiments. An investigator puts forward a hypothesis about how something works and then designs an experiment to test this idea, yielding results that either support or refute it. Then, either the scientist himself or a rival comes up with a critique of the conclusions, presents an alternative hypothesis and further experiments for testing the idea, and so on, back and forth like a tennis match, during which time the explanation of the phenomenon in question gets closer and closer to the truth. From this point of view, the Galvani/Volta debate was excellent science because each experiment was carefully designed to precisely test the suggested hypothesis. The other thing that almost invariably happens in this kind of situation is that valuable information is obtained about "stuff in general," that is, things that weren't necessarily involved in directly testing the hypothesis. In this case we can point to Volta's discovery of metal electricity, something he proposed as an alternative explanation for Galvani's results and that he then developed further, ultimately leading to the invention of the battery. It is hard to imagine how Galvani and Volta could have produced these advances at the time when they lived without exploiting animal preparations in the way they did. And really this is what animal research is best at—finding out basic things about biology. Harvey had found out how the cardiovascular system worked, and now Galvani had discovered that electricity had an important role to play in the control of nerve and muscle function. Although there are clearly differences between frogs or electric fish and humans, some basic biological processes transcend these differences. The cell membrane of a frog or mouse cell is constructed in a very similar fashion to a human cell, electricity is involved in nerve conduction in humans in a very similar way as in a frog or mouse, and so on. There are limits, of course, but it is certainly true that in some respects animals can be reasonable analogs of humans; it just depends on what you happen to be investigating. Where things often become more difficult, as we will discuss (see Chapter 4), is when one tries to use animals for translating therapeutic findings to humans, something that rarely works because of important differences that also exist between animals and humans. Moreover, just because animal research can yield important information about biology, this doesn't mean that we should be doing it. That is a question of ethics.

Another interesting facet of Haller's, Galvani's, and Volta's experiments was their increasing use of repetition. At some point in the history of science, repetition would be codified into the discipline of probability and statistics, which would generate mathematical tools for testing whether experimental results were reliable or not. In the seventeenth century Sir Francis Bacon's scientific manifesto had pointed out that repetition was an important factor when making measurements and generating reliable data. Of course, in biology, greater repetition means increasing use of animals. Eventually, the biological sciences would become completely dependent on statistical tests to determine the validity of experimental measurements, and more and more animals would need to be sacrificed to the cause.

French Physiology in the Nineteenth Century

By the dawn of the nineteenth century, along with the growth of the middle classes, more people were becoming engaged in carrying out scientific experiments, and some people, like Robert Hooke, who became the chief "demonstrator" for the Royal Society in London in 1660, were making a profession out of it. It was no longer necessarily an occupation just for wealthy gentlemen but for different types of individuals who specialized in these pursuits. Soon enough, somebody came up with a word to describe these people. Pursuers of what had been called "natural philosophy" became "scientists." This new profession developed at different rates depending on the country in question. In a more patrician society like England, science remained a "rich man's plaything" for quite some time, but in countries with a more revolutionary frame of mind, such as France, the professional scientist became a much more common individual. The way animal experimentation developed in the two countries also differed greatly. England, with Victoria as its animal-loving monarch, was to see the birth of the first political forces aimed at trying to protect animals from abuse (see Chapter 7). But France was to see the rise of some famous scientific figures who became legendary for the extraordinary cruelty with which they set about their experimental endeavors. This split was encouraged by, among other things, the historical love/hate relationship between the British and the French, in which the British cast the French experimental physiologists as barbaric savages and tried to intervene in their research, and the French told the British to mind their own business. Nevertheless, it was the French way of doing biological research and of using animals that would ultimately become the model for science worldwide, including in Britain.

The two scientists who most typified the French approach to biology were Francois Magendie and his pupil Claude Bernard.[19] Both were extremely influential but also extremely divisive figures in terms of what they achieved and how they achieved it. During the late eighteenth and early nineteenth centuries, the revolutionary violence that gripped France resulted in large amounts of human and animal material (particularly horses) that could be potentially studied by medical anatomists. A great deal of information resulted from this work, establishing France as the center of research in this area. However, Magendie, who held the chair of medicine at the College of France in Paris, was opposed to this line of investigation. How could research on cadavers tell you anything about life, the major thing that was clearly absent under these circumstances? On the other hand, if one began with a live creature, there would be many things one could investigate, and as a bonus, as the life of the animal drained away, analysis of the manner of its death might also lead to insights about what was eventually lost. As one of his students Julien Legallois wrote, "Experiments on living animals are among the greatest lights of physiology. There is an infinity between the dead animal and the most feebly living animal." In other words, an animal was either alive or dead. Nowadays we don't think in such precise terms, and the definition of death is constantly under scientific, medical, and legal review (see Chapter 6). Moreover, as far as animal pain was concerned, Magendie thought that it was a natural expression of the living process, so efforts to suppress it were simply nonscientific. During the early part of his career, Magendie, like Harvey and Galvani before him, set up a laboratory at home and ran courses on experimental physiology for budding medics and scientists. Vivisection was the core of his experimental approach, dogs being his favorite subjects. And apparently Magendie was a very skillful dissector. Nothing was beyond his purview; no surgical intervention using a live animal, however excruciating, was beyond the pale as far as he was concerned. What is more, much of what Magendie did would be considered excellent science even today. He would present a clear hypothesis and test it rigorously, exactly as the US National Institutes of Health (NIH) would have you do if you are to successfully win a research grant. For example, he asked what were the functions of the two nerve roots that came out of the spinal cord? Cut one root and you were left with an animal that could feel pain but no longer walk; cut the other and you were left with an animal that could still walk but could not feel pain. Magendie's description of the sensory and motor functions of the spinal roots is considered to be a landmark in the history of physiology. The obviously excruciating pain he induced in his animal subjects was simply not a consideration. Because Magendie's work elicited strong negative reactions in some observers, there are many descriptions in the literature of the kind of thing he did. One eyewitness account will suffice.

> Magendie, alas! performed experimentation in public, and sadly too often at the College de France. I remember once, amongst other instances, the case of a poor dog, the roots of whose spinal nerves he was about to expose. Twice did the dog, all bloody and mutilated, escape from the implacable knife; and twice did I see him put his forepaws around Magendie's neck and lick his face. I confess—laugh vivisectors if you please—that I could not bear this sight.

Apparently, that is exactly what Magendie used to do—laugh. He was frequently reported to make fun of his animal subjects as he wielded his knife. Worrying about such things was not for the strong and forthright, the pure of spirit, the soldiers who would march at the frontlines of science into a glorious future. Magendie was an emotional "ubermensch," to use Nietzsche's descriptor, placing him and other scientists who were his acolytes above normal humans whose emotions rendered them weak and unable to perform important scientific work. For Magendie, his demonstrations were pure theater. Here was the original "theater of cruelty," which, as subsequently described by Antonin Artaud, demanded that "this cruelty, which will be bloody when necessary but not systematically so, can thus be identified with a kind of severe moral purity which is not afraid to pay life the price it must be paid."

Magendie was lauded for his scientific work in France. In England, however, his kind of approach was not particularly palatable to the Romantics, who represented a powerful reaction to Enlightenment thinking. Science and scientists who had reneged on their moral responsibilities would not necessarily lead us anywhere good, as Mary Shelley so ably pointed out.

Magendie's experimental program was inherited by his most able student, the equally divisive Claude Bernard. Bernard began his career studying medicine, which he never practiced, and became Magendie's assistant. Following Magendie's death Bernard took his place as professor at the College de France. He needed funding to carry out his program of animal experimentation, so he married a wealthy woman. Unfortunately, she eventually became so disgusted by Bernard's animal experiments that she left her husband and joined the burgeoning antivivisection movement. Bernard's experimental methods closely followed those of Magendie's, but if anything, Bernard was an even more brilliant and forceful scientist. He made many discoveries concerning subjects like the physiology of digestion and the pharmacological properties of the Amazonian Indian arrow poison curare, but perhaps his most important contribution was a theoretical one—the idea of homeostasis, or *milieu interieur*. Bernard considered the body to be a series of connected machines that had particular mechanistic functions that one could study in isolation. However, the phenomenon of life depended on all these elements constantly interacting with each other in an appropriate manner utilizing an array of feedback and feedforward circuits. These could be neural or what would ultimately prove to be endocrine in nature. When something altered the beating of the heart and the movement of the blood, other tissues such as the gut, the liver, and the kidneys had to alter their functions appropriately, making the overall operation of a living animal analogous to something like a steam engine. This cooperative association of organ systems was what really constituted life. All of the phenomena he studied were subject to rigorous experimentation, and like Magendie, Bernard called for scientists to act as heroes. "The physiologist is not an ordinary man: he is a scientist, possessed and absorbed by the scientific idea that he pursues. He doesn't hear the cries of animals, he does not see their flowing blood, he sees nothing but his idea, and is aware of nothing but organisms which conceal from him the problems he is wishing to resolve." Here again, the scientist as Nietzschean superman was very much the model.

According to Bernard, the scientist might view his animal subjects as more of an impediment trying to prevent him from reaching the truth than as an aid to his endeavors. The role of the scientist was to conquer Nature, and impediments needed to be swept aside. This was science as war. No wonder his cruelty was legendary. The fact is, however, that the kinds of statements made by Magendie and Bernard, requiring a sense of intellectual rigor, purity and emotional detachment, are still frequently made by scientists today to justify what they are doing, even if the precise verbiage has changed somewhat. Nowadays, of course, there are anesthetics that can prevent some of the pain associated with vivisection. However, this creates an entirely different problem, what we now call "survival surgery" (see Chapter 7). At least in the hands of Magendie or Bernard an animal might finally escape its agony in death. Nowadays even that is denied to it as, following recovery, an animal is forced to live out its life forever suffering from whatever has been inflicted upon it. The latest tweak on all of this is to manipulate the genomes of animals so that they are inevitably destined to develop horribly debilitating diseases and suffer throughout their lives. We will discuss these issues further when we analyze the utility of animal models for research purposes (see Chapter 4).

Magendie and Bernard took the vivisection-as-experimental paradigm about as far as it could go. The types of experiments I observed in the seminar described in the Chapter 1 are clearly the descendants of their approach. But as the nineteenth century progressed, things were starting to change. There was another type of idea that was being developed in Germany at around the same time in laboratories, such as that of Johannes Muller in Berlin.[20] This involved an even more basic and totally materialist approach to the phenomenon of life: an approach based on cell biology and biochemistry. The concept of the cell and its structure as an experimental paradigm for understanding living organisms was utilized very effectively by Muller's students, such as Theodor Schwann, Robert Remak, and Rudolf Virchow. Galvani's concepts on the role of bioelectricity would be further advanced as an aspect of cell biology by Muller's student Emil Dubois-Reymond and his own students, Ludimar Hermann and Julius Bernstein. The new united German state prioritized science and gave it ample support, and soon German science would become the dominant voice in biology and medicine for about one hundred years.

Bernard's influence was so widespread particularly because he also wrote the *Introduction to the Study of Experimental Medicine,* a classic text that clearly set out his scientific worldview and detailed experimental approach. Naturally, this wasn't well received in England, where antivivisectionist feelings were starting to gain some real traction. But for scientists, even in England, the kind of work being performed by Magendie and Bernard was just too seductive. It really was scientific in an appropriately modern sense that relied only on clear reasoning, hypothesis, and experiment and did away completely with any hint of the metaphysical. The use of rigorous scientific methodology could enable every scientist's ultimate dream of establishing total control over Nature. Soon enough these ideas would also obtain a foothold in England. In 1873 John Burdon Sanderson, a student of Bernard's, and now a professor of physiology in London, published his *Handbook for the Physiological Laboratory,* a primer on how to perform animal experiments a la Bernard for teaching and research purposes. All these studies were to be carried out without the use of anesthetics, which were, by this time, in common usage. When asked why he hadn't mentioned this, Sanderson replied that he assumed everybody used them, so he felt he didn't have to mention it, although that clearly wasn't the case. Not mentioning the use of anesthetics presumably wasn't, to use Oscar Wilde's phrase, "merely carelessness." More likely it was just callousness. When one of the contributors to the book was asked about his attitude toward pain in animals subjected to the experiments he had described, he replied that he had "no regard for it at all." Like his vivisectionist predecessors, Sanderson was an excellent scientist and made several important contributions to physiology, but his book set him on a direct collision course with the British antivivisectionist movement, an interesting story that we shall discuss later (see Chapter 7). His textbook was extremely influential, and despite vociferous opposition, the British experimental tradition, including the use of animals, was destined to track that established on the European continent from that point on.

Thus, by the turn of the twentieth century, the animal machine set in motion by Descartes had become the intellectual and practical model directing the majority of biomedical research, and the approach worked—as science. The advances made by many of the scientists we have discussed during the period from the seventeenth to twentieth centuries form the basis of much of what we still believe in the twenty-first century. Scientists used animals to great effect when performing their investigations. Many of the individuals responsible for this work are heralded as great scientists today—models of rigorous scientific thinking and experimentation—and the truth of this cannot be denied. However, it is also just as true that the things they achieved were the result of terrible cruelty to animals. How, then, do we judge individuals like Magendie and Bernard—as great men or sadistic monsters, the same question we asked about Galen and Harvey? This is the quandary we still find ourselves facing today. But many things have changed since Bernard's time, and the context in which biomedical research is carried out nowadays is very different. Whatever Bernard and others achieved, the important question we need to ask ourselves is whether their approach is still valid today or whether we can obtain equally important information in the service of humanity that doesn't produce so many tragic victims—a question we will now begin to address.

Notes

1. Richard E. Rubenstein, *Aristotle's Children: How Christians, Muslims, and Jews Rediscovered Ancient Wisdom and Illuminated the Dark Ages* (Prince Frederick, MD: RB Large Print, 2003); James Hannam, *The Genesis of Science: How the Christian Middle Ages Launched the Scientific Revolution* (Washington, DC: Regnery Pub, 2011).
2. M. H. Green, "Gloriosissimus Galienus: Galen and Galenic Writings in the Eleventh- and Twelfth-Century Latin West," in *Brill's Companion to the Reception of Galen,* ed. Petros Bouras-Vallianatos and Barbara Zipser (Leiden: Brill 2019), 319–342, doi:10.1163/9789004394353_019; Rubenstein, *Aristotle's Children;* Hannam, *The Genesis of Science.*
3. W. C. Aird, "Discovery of the Cardiovascular System: From Galen to William Harvey: Discovery of the Cardiovascular System," *Journal of Thrombosis and Haemostasis* 9 (July 2011): 118–129, doi:10.1111/j.1538-7836.2011.04312.x.
4. Thomas Wright, *Circulation: William Harvey's Revolutionary Idea* (London: Chatto & Windus, 2012).
5. Aird, "Discovery of the Cardiovascular System"; Wright, *Circulation.*
6. Wright, *Circulation.*
7. Aird, "Discovery of the Cardiovascular System."
8. A similar suggestion had also been made by the Arab doctor Ibn al-Nafis in the thirteenth century, but his writings had not been disseminated in Western Europe and so scientists such as Harvey did not know about them. See John B. West, "Ibn Al-Nafis,

the Pulmonary Circulation, and the Islamic Golden Age," *Journal of Applied Physiology* 105, no. 6 (December 2008): 1877–1880, doi:10.1152/japplphysiol.91171.2008.

9. Jessica Riskin, *The Restless Clock: A History of the Centuries-Long Argument over What Makes Living Things Tick* (Chicago: University of Chicago Press, 2016); Luke Morgan, *Nature as Model: Salomon de Caus and Early Seventeenth-Century Landscape Design*, Penn Studies in Landscape Architecture (Philadelphia: University of Pennsylvania Press, 2007).

10. Riskin, *The Restless Clock*.

11. Michael Hunter, *Boyle: Between God and Science* (New Haven, CT: Yale University Press, 2010).

12. Lisa Jardine, *The Curious Life of Robert Hooke: The Man Who Measured London* (New York: Perennial, 2005).

13. Hunter, *Boyle*.

14. I didn't get the joke either. Here is what Wikipedia says about it: "Batrachomyomachia," https://en.wikipedia.org/wiki/Batrachomyomachia, accessed July 5, 2023.

15. Ibtisam Azaiza, Varda Bar, and Igal Galili, "Learning Electricity in Elementary School," *International Journal of Science and Mathematics Education* 4, no. 1 (March 2006): 45–71, doi:10.1007/s10763-004-6826-9.

16. For a complete discussion of Galvani and Volta's experiments see Marco Piccolino and Marco Bresadola, *Shocking Frogs: Galvani, Volta, and the Electric Origins of Neuroscience* (Oxford: Oxford University Press, 2013).

17. Stanley Finger and Marco Piccolino, *The Shocking History of Electric Fishes: From Ancient Epochs to the Birth of Modern Neurophysiology* (New York: Oxford University Press, 2011).

18. Finger and Piccolino; *The Shocking History of Electric Fishes*; Piccolino and Bresadola, *Shocking Frogs*.

19. For a review of the history of animal research discussed in this chapter see Anita Guerrini, *Experimenting with Humans and Animals: From Galen to Animal Rights*, Johns Hopkins Introductory Studies in the History of Science (Baltimore, MD: Johns Hopkins University Press, 2003); James Montrose Duncan Olmsted and E. Harris, *Claude Bernard and the Experimental Method in Medicine* (Chicago: H. Schuman, 1952).

20. Laura Otis, *Müller's Lab* (Oxford; New York: Oxford University Press, 2007).

4

Mapping Humanity

Nothing will come of nothing.
—*King Lear* Act 1i, William Shakespeare

As we have seen, since Galen's time in ancient Rome, much of the research designed to uncover the secrets of the human body has relied on our ability to make useful analogies with animals. According to the nineteenth-century English philosopher of science John Herschel, "If the analogy of two phenomena be very close and striking, while, at the same time, the cause of one is very obvious, it becomes scarcely possible to refuse to admit the action of an analogous cause in the other, though not so obvious in itself." Animals have been used as stand-ins or metaphors, particularly when we think it wouldn't be ethical to perform experiments directly on human beings. They have acted as maps, helping us find our way around the human body when we can't do so directly. But what systems make the best maps or models, and how useful are they in practice? The writer Jorge Luis Borges wrote a short story about this problem called *On Exactitude in Science*. It is very short indeed, being only one paragraph long, so here is the entire text.

In that Empire, the Art of Cartography attained such Perfection that the map of a single Province occupied the entirety of a City, and the map of the Empire, the entirety of a Province. In time, those Unconscionable Maps no longer satisfied, and the Cartographers Guilds struck a Map of the Empire whose size was that of the Empire, and which coincided point for point with it. The following Generations, who were not so fond of the Study of Cartography as their Forebears had been, saw that that vast Map was Useless, and not without some Pitilessness was it, that they delivered it up to the Inclemencies of Sun and Winters. In the Deserts of the West, still today, there are Tattered Ruins of that Map, inhabited by Animals and Beggars; in all the Land there is no other Relic of the Disciplines of Geography.[1]

The Rise and Fall of Animal Experimentation. Richard J. Miller, Oxford University Press. © Richard J. Miller 2023.
DOI: 10.1093/oso/9780197665756.003.0004

As Borges pointed out, no map or model will tell us everything about the territory it is supposed to represent unless it is actually the territory itself, and the same is true when considering experimental models of human beings. What do they tell us and what is missing? As the statistician George Box once said, "All models are wrong. The question is are they useful?" When modern science began to advance once more in the seventeenth century, Sir Francis Bacon commented on the potential use of animals to model aspects of human biology, writing in his *De Augmentis Scientiarium* (1623), "Wherefore that utility may be considered as well as humanity, the anatomy of the living subject is not to be relinquished altogether . . . since it may be well discharged by the dissection of a beast alive, which, notwithstanding the dissimilitude of their parts to human, may with the help of a little judgment, sufficiently satisfy this inquiry." Unfortunately, the fact that animal models of humans are invariably lacking in certain key respects is something that is frequently ignored these days by scientists who seem to be ignorant of Wittgenstein's advice, "Whereof one cannot speak, thereof one must be silent."

In 1739, a French natural philosopher proposed a scheme for testing new medical therapies. The appropriately named Claude-Nicolas Le Cat suggested constructing an automaton that displayed many of the important functions of a human being. It would have "all the operations of a living man . . . the circulation of the blood, the movement of the heart, the play of the lungs, the swallowing of food, its digestion, the evacuations, the filling of the blood vessels and their depletion by bleeding . . . even speech and the articulation of words."[2] All of this, of course, was in keeping with Cartesian ideas circulating at the time that the bodies of humans and animals were essentially mechanical. So why not make a machine that reflected these realities and use it for developing and testing medicines and surgical procedures prior to using them on humans?

This is an important idea and is the basis for much of the biomedical research performed today. Rather than developing new therapies by testing them on humans, why not use a human analog instead? Devices like these would have a number of advantages. They might be cheap to make and easy to use so that one could try out multiple iterations of a proposed therapy before choosing the best alternative, thereby avoiding any untoward effects on human subjects during the process. Le Cat's ideas were extraordinarily ahead of their time and presaged current attempts to use computer simulations and other nonorganic paradigms for testing the potential of new drugs or other therapies for human diseases. Of course, there is also another way of achieving these aims; following Bacon's suggestion, why not use animals as models of humans and practice on them instead? After all, there are similarities between humans and animals and, as we have seen, according to Cartesian ideas, animals may not be as sensitive to interventions that would be painful or unpleasant to humans. Moreover, animals don't have the same moral status as humans and, according to Christian ideas, were put on Earth to serve human needs in any way we might deem appropriate. And of course, it is Christian morality that has always provided the lens through which the Western countries who developed modern science see the world.

Arguments like these have become the basic credo for all animal-based contemporary biomedical research, and hundreds of millions of animals are used for such purposes every year—around 110 million at the most recent count.[3] Given the pervasiveness of this approach, the enormous numbers of animals involved, and the billions of dollars invested in this type of experimentation every year, we need to take a careful look at it, understand why scientists use animals as models, and importantly, ask if and when they are actually useful.

As we have seen, animal experimentation started in the time of the ancient Greeks and developed over the next several thousand years until it reached its contemporary form. This development occurred in two major bursts: the first took place in antiquity, and the second, which started in the seventeenth century, continues today. We have also seen that the use of at least some animal experimentation in the hands of great scientists like William Harvey or Claude Bernard enabled humanity to make extraordinary advances in our understanding of biology and medicine. But right from the start, the use of animals went hand in hand with terrible cruelty. When it came to animal experimentation, scientists seemed devoid of empathy for the creatures they dissected or vivisected. Our current use of animals reflects many features of this past development. In most important respects, the investigations carried out by William Harvey or Claude Bernard look a lot like investigations that might be carried out today; and the seminar I described in Chapter 1 is clearly a descendent of this approach. Of course, some things have changed, and these things are extremely important, particularly the use of various drugs and anesthetics during animal surgery. There have also been numerous technical advances that determine precisely what types of experiments can be carried out in principle, and what types of things can be measured. Scientists working in biomedical research today will surely tell you that animal research is absolutely necessary in order to make progress and that, unfortunately, it is just the price we have to pay if we want to cure humanity of terrible diseases; there really isn't any alternative, even today. I cannot tell you how many times I have heard this said, although in my opinion, it is usually simply a reflexive response and the researchers in question have probably never given the subject any real thought at all. But, as it turns out, there are an ever-increasing number of alternatives to animal research these days. Big changes are on the horizon.

Basic and Applied Research

What exactly does a researcher hope to achieve when performing an experiment on an animal? We can probably divide today's research into two main types—basic and applied. Basic research is designed to add to the corpus of human knowledge about the natural world. Research of this type really harkens back to Aristotle, who was generally interested in understanding Nature writ large. The Greeks thought that it should be a major goal of humanity to understand Nature as perfectly as possible as well as the place of humanity in the scheme of things. This is still an opinion that would be held by most people today. Science is viewed as an important aspect of human culture. A general knowledge of how the world works is essential for any individual who wants to be a well-informed, sophisticated, cultivated, and (hopefully) useful member of society. But we don't just wish to understand Nature. We also desire to dominate her so that we can bend her to our will and shape her according to our vision of ourselves. It is this aspect of human curiosity that will particularly appeal to scientists. A scientist wants not only to understand Nature but also to control and change her, hopefully for the ultimate benefit of humankind. Unfortunately, the idea that biomedical research should be directed toward improving the human condition and that nothing is more important than this aim allows scientists to do virtually anything they want with a clear conscience, leading to a kind of scientific antinomianism where it is impossible to call them to account for either their intellectual or moral failings.

Generally speaking, good basic biomedical research isn't necessarily about curing a particular disease or medical problem; it's about adding to the corpus of knowledge that allows humanity to better understand itself and the world in general. Nevertheless, this kind of knowledge often does ultimately contribute to treating diseases or solving other human-related problems.

As Galvani demonstrated, the fact that nerves and muscles use electricity to carry out their physiologically important functions is important basic information about how the body works. But these ideas are also some of the cardinal foundational concepts that underlie our understanding of neurology and cardiology and are the basis of many of the therapeutic procedures these disciplines have developed in the world of clinical medicine.

The choice of animal used in an experiment will naturally depend on what the experiment is designed to investigate. Let us say you are a basic researcher, and you are interested in snakes. Finding out about snakes is your passion and you want to conduct research on them come what may. If that is the case, then the experiments you perform will, by necessity, concentrate on snakes. Perhaps you are interested in the differences in the cardiovascular systems of various types of snakes—something that we might call "comparative physiology." Well, then, there is no real way around performing studies on the snakes in question. There would be little point in performing experiments on mice, wombats, eagles, or even humans to answer the questions that interest you. In your experiments snakes are being used as experimental models of themselves. These experiments have the potential to answer the questions you are interested in. So, nothing is missing from your animal model.

On the other hand, the main type of biomedical research carried out today is known as applied translational research. This means that the research that you do usually has to be seen as being directly relevant to humans and capable of potentially answering some question about a human-related problem. Most of this research in the United States, for example, is funded by government agencies such as the National Institutes of Health (NIH, or equivalent bodies in other countries), whose stated aims are to cure human diseases and improve the lot of humanity. It is therefore not surprising that if you want to obtain research funding from the NIH, they will want to know how the money you receive is going to help it to achieve its stated goals.

Some of the work the NIH funds, which is generally carried out by medical doctors, is concerned with testing novel therapies on the human population. This is known as clinical research and often involves performing what are known as clinical trials on selected groups of humans. The relevance of this research to humans is therefore quite straightforward. However, if a basic researcher wants to carry out NIH-funded research, the proposed experiments will often involve the use of animals as models for humans. In other words, as opposed to the snake researcher we discussed above, applied research isn't ultimately interested in the biology of animals per se but only what it can potentially tell us about the biology of humans. The researcher will have to explain to the NIH how their research is going to be relevant to some human condition. To use the terminology currently in vogue, the researcher will have to explain how their research can be "translated," that is, "be directly relevant to" humans. It may be that the researcher can think of a reason why experiments using snakes, wombats, jellyfish, or eagles might fulfill this goal, but the NIH will have to be convinced of this fact if they are going to hand over any money.

An interesting question, therefore, is what factors go toward making an animal model where the results can be translated to humans in a meaningful manner or, indeed, whether such a thing is even possible? Perhaps some might argue it isn't, and if we want to understand humans, as Wittgenstein (again) said, "The human body is the best picture of the human soul."

The Rise of the Pharmaceutical Industry

Prior to doing so, however, we should spend a little time introducing another important institution that is perhaps the greatest consumer of research animals in today's world: the pharmaceutical/cosmetics industry, which uses up a large chunk of the 110 million or more vertebrate animals used by biomedical researchers every year.[4]

In contrast to the types of research we have been mostly concerned with, the pharmaceutical industry is the new kid on the block. By this I don't mean that pharmacology, the science of drugs and how they work, is new. Far from it. Pharmacology is one of the oldest of all scientific endeavors, and writings about drugs and their uses go back to sources like the Ebers papyrus (ancient Egypt) or the pharmacopeias of ancient Greece, China, and India.

Of course, originally all these medicines had to be sourced from plants and animals. Some of them worked very well. We should consider things like quinine (malaria), digoxin (heart), and morphine (pain), to mention just a few famous examples that humans have discovered from the treasure trove of medicines that have come directly from Nature. Natural products were often used in a sophisticated manner. Some ancient remedies were extremely complex concoctions of different substances. We know that Mithridates VI of Pontus (circa 120–63 BCE), one of the most renowned pharmacologists of the ancient world, developed a "universal antidote" which contained more than sixty individual natural ingredients. Following his death, Roman physicians spoke of an *Antidotum Mithridaticum* (mithridatium) formula, which became a sought-after pharmaceutical preparation until the end of the Middle Ages,[5] used for treating a variety of diseases.

Moreover, it is also clear that even in ancient times animals were sometimes used for testing the effectiveness and potential toxicity of drugs. In *De Theriaca ad Pisonem*, Galen described how he took roosters and divided them into two groups: one group was given mithridatium and the other group was not. Then Galen brought both groups into contact with vipers; he observed that the roosters who had not been given mithridatium died immediately after being bitten, whereas those who had been given the drug survived. One might conclude from this experiment that taking mithridatium would also protect humans, as opposed to roosters, from the effects of being bitten by a viper, and this is essentially the type of experiment that might be conducted by a pharmacologist today.

Drugs have been an essential part of human existence from time immemorial. However, originally all these substances were the result of Nature's bounty. Drugs were widely used, but humans weren't in the business of coming up with anything new that wasn't already provided for them by Nature. And things remained that way until the nineteenth century, when something momentous was invented. This was called "organic chemistry." Organic chemistry concerns the chemistry of the carbon atom, which forms the basis of all living things, and this knowledge allowed humans to not only start isolating but also changing the chemicals they found in Nature. This was the gateway to inventing new drugs, using chemistry to alter the fabric of Nature at will, and helping to fulfill the scientific dream of controlling the natural world.

The very first drug companies began their operations in the nineteenth century. To begin with most of them were dye-making companies.[6] They made synthetic dyes from the organic chemical aniline, which was widely available as a by-product from the coal mining industry that was powering the incipient Industrial Revolution. Quite early on many of these companies started to ask whether their new aniline-based chemical products might have additional properties. The German company Bayer, one of the very first drug companies, showed that some of their products were able to reduce fevers. For example, the drug acetaminophen (Tylenol) was the result of this line of investigation. Bayer also found that they could modify natural products with newly invented chemical reactions. Acetylation of salicylic acid derived from willow bark produced acetylsalicylic acid, an excellent drug for reducing fevers that they named aspirin. They tried the same reaction on morphine isolated from poppies and produced diacetylmorphine, which they named heroin. Bayer thought heroin was an excellent substitute for morphine because it wasn't addictive (!).[7] The chemical floodgates were now open, and drug companies started to produce thousands of new products that they were anxious to test as drugs.

Their task was about to get even more interesting because of another major nineteenth-century discovery. The work of the likes of Louis Pasteur and Robert Koch demonstrated that many infectious diseases were caused by microbes or "germs." The germ theory of disease, therefore, created a new target for drugs: kill the microbes and cure the disease! But how could this be achieved? The answer was provided by Paul Ehrlich, one of the most important drug developers in history.

One of the greatest plagues of the nineteenth century was the sexually transmitted disease known as syphilis. It had a status at the time that was somewhat akin to HIV-1 in the twentieth century. Not only was syphilis a terrible disease with frequently fatal consequences, but also it came with a great deal of social baggage. There was very little that could be done about it. Historically, patients were chronically treated with toxic metals such as mercury and arsenic, leading some wit to quip about syphilis, "One night with Venus and a lifetime with Mercury." This treatment was known as iatrochemistry, an idea attributed to the sixteenth-century alchemist Paracelsus,[8] which may have helped somewhat but was so poisonous that the cure was often worse than the disease. But it was the treatment of syphilis that would really launch the pharmaceutical industry in its modern incarnation.

While working as a student, Ehrlich had become interested in the science of histology, that is, how to use dye staining to identify different types of cells and materials in tissue sections. The science of histology is based on the fact that many dyes stain tissues differently. One dye might be better for staining nerves, one might be better for muscle, and so on. This gave Ehrlich a brilliant idea. Perhaps chemicals could be made to stick to different types of infectious microbes, rather than the tissues of the host, and then kill the microbial cells selectively. These "magic bullets" would be the medical equivalents of intercontinental ballistic missiles homing in on their targets and destroying them. Later, the same principle would be applied to cancer cells. The idea came to be known as "chemotherapy."[9] How, then, could one create a magic bullet that would target the organism responsible for syphilis?

The bacterium responsible for the disease had been identified as the spirochete *Treponema pallidum*. Ehrlich realized something interesting. Arsenic had been used for many years for treating numerous diseases including syphilis, but of course it was also extremely poisonous. The organic chemical aniline had been used to kick start the entire pharmaceutical industry, and in 1869 a French scientist named Antoine Beauchamp had the idea that he might be able to make arsenic less toxic, and therefore more useful, by reacting it with aniline.[10] The resulting material, known as Atoxyl, found little practical application until, in 1905, scientists in England found that it might be useful for treating African sleeping sickness, a disease also caused by an infectious micro-organism.

Ehrlich thought that Atoxyl, which was by no means completely free of toxicity, might be a suitable starting point for creating a more specific arsenic-containing drug for treating syphilis. To achieve this aim, Ehrlich formed what was in reality a mini drug company consisting of a chemist, Alfred Bertheim, and two pharmacologists, Ehrlich and his colleague Sahachiro Hata. Hata was able to successfully infect mice and rabbits with *T. pallidum*. This, then, was their animal model for the human disease. Starting with Atoxyl, Bertheim used organic chemistry to make small chemical modifications to the drug and gave each substance to Hata and Ehrlich to test on infected animals. Would any of the compounds cure them without showing untoward toxic side effects? Bertheim would be informed of each result and would then make further compounds in the hope that they would have improved properties. In this way they would eventually zero in on the winner—the molecule that would best cure the disease without poisoning their animal subjects. Ultimately, the team came up with "compound 606," which had the best combination of properties. Next it was time to test the drug on humans. Success! The drug worked very well as a cure for human syphilis; patients were frequently completely relieved of their disease. The drug, which they named arsphenamine, was marketed in 1910 under the name Salvarsan.[11] The procedural flow chart used by Ehrlich and his colleagues was adopted by drug companies as their main discovery plan, and it has changed little in terms of the basic steps employed, even today. First, select a lead compound and find an assay for testing it; this is frequently an animal model of the disease. Then, make compound after compound, testing as you go, and refine its chemistry until the ideal drug has been created. It's like a conveyer belt, making new drugs and then giving them to animals to test their effects. Even in the case of Salvarsan, thousands of animals were used for drug testing, and today in large pharmaceutical companies with multiple drug development programs being carried out simultaneously, the number of animals used has skyrocketed into the hundreds of millions.

Indeed, when it comes to curing infectious diseases, Salvarsan was only the tip of the iceberg. In 1935, sulfanilamide and other sulfa drugs were introduced for combating numerous infectious diseases, and then finally came the biggest breakthrough of all—the discovery of antibiotics during the Second World War. The intense efforts made to develop these new wonder drugs used millions upon millions of infected animals as living Petri dishes for assaying new antibiotics. Vaccines too would be an enormously important group of biomedical products and would also take their toll on animals. It has been said that the development of polio vaccines in the 1950s decimated the monkey population in India by a third—an incredible number.[12] In summary, the rise of what we now call "medicinal chemistry" produced an unprecedented number of new chemicals that needed to be evaluated by the pharmaceutical industry for their potential therapeutic effects, and by the 1960s, testing these substances on live animal models of human diseases or on isolated tissues provided by animals for bioassays formed the backbone of pharmaceutical research throughout the world.

However, the invention and preclinical testing of new drugs, vaccines, and other products of this type (termed "preclinical research") is not the only reason pharmaceutical companies use animals. Another major reason is to comply with federal regulatory safety laws. For most of human history, drugs were freely available to anybody who wanted them for any purpose. If Thomas De Quincy wanted some opium, he would just walk into an apothecary and buy some. There were no restrictions on him doing so. There were also numerous medical concoctions for sale in the nineteenth century, generally known as nostrums or patent medicines, which claimed to do all kinds of miraculous things, although nobody knew what was in them. Perhaps some dried snakeskin mixed up with salt and sulfur would be sold as "Dr. X's Wonderful Cure for Rheumatism!" (or perhaps as Tono-Bungay, as in H. G. Wells's novel of the same name, which satirizes the use of patent medicines).

Some of these concoctions were extremely dangerous. Medicines like "Mrs. Winslow's Soothing Syrup" were sold to mothers who had to work in factories during the Industrial Revolution. They had to leave their children at home, and it was useful to have something that would make the children sleepy and easy to look after by nannies or elderly relatives. The syrup was basically a tincture of opium, so the children who took it were usually very sleepy indeed.[5] By 1900 governments had started to become interested in investigating the potential toxicity of medicines, food additives, and environmental toxins. In the United States, for example, Dr. Harvey W. Wiley of the U.S. Department of Agriculture (USDA) Division of Chemistry organized his famous Poison Squad to investigate these matters. Twelve USDA employees acted as human subjects between 1902 and 1904 in testing the safety of boric acid, salicylates, sulfurous acid, benzoic acid, and formaldehyde.

The Poison Squad even had their own song:[13]

O we're the merriest herd of hulks that ever the world has seen;
We don't shy off from your rough on rats or even from Paris green:
We're on the hunt for a toxic dope that's certain to kiss, sans fail,
But 'tis a tricky, elusive thing and knows we are on its trail'
For all the things that could kill we've downed in many a gruesome wad
And still we're gaining a pound a day, for we are the Pizen Squad.

Wiley recognized that the results from using animals might not be representative of what potential toxins might do to humans, and the relative benefits of using animals or human volunteers were widely discussed.

Gradually, laws started to be written that were designed to regulate drug use. These included requiring prescriptions for some drugs, clearly displaying descriptions of ingredients on drug bottles so that people could see what was in them, and international treaties to restrict the sale of very addictive drugs like opiates and cocaine.

Laws were also put into place requiring drug companies to show that their products were both safe and effective. There were several dramatic incidents that forced politicians in the United States and other countries to pass such laws. For example, in 1937, an American company named SE Massengill marketed a preparation ("elixir") of sulfanilamide in diethylene glycol (DEG) together with raspberry flavoring to make it more palatable. Unfortunately, the DEG was extremely toxic and over a hundred people who took the medicine died.[14] There was a huge hue and cry triggered by these events, forcing the US government to take a careful look at the way drugs were developed and marketed, and in 1938, Congress enacted the Food, Drug, and Cosmetics Act. These regulations required pharmaceutical companies to carry out a number of tests using animals to demonstrate the safety and effectiveness of their products prior to presenting their data to the Food and Drug Administration (FDA) for potential approval and marketing to the public.

Subsequently, following the discovery of widespread toxic effects of the drug thalidomide when used by pregnant women, the Kefhauver-Harris amendment of 1962 added further regulations, resulting in a very large number of tests that needed to be carried out by the pharmaceutical industry.

These regulations are complicated but cover numerous experiments designed to establish the safety and effectiveness of any new drug. For example, new drugs need to be tested in several different animal species and on both sexes. Tests also need to cover both acute and chronic administration of drugs. Other tests are required for examining the effects of a drug on a variety of physiological systems including reproduction and embryological development, as well as testing for potential cardiac, neurological, respiratory, kidney, and gastrointestinal problems. Companies must also carry out procedures to establish the ADME (absorption, distribution, metabolism, and excretion) profile of a new drug. To these tests we can add toxicology, resulting in ADMET.[15] Overall, these studies are supposed to provide a reliable description of what a drug does to an animal and what an animal does to a drug, the idea being that if a drug is toxic to animals, it will also be toxic to humans, providing a potential safety barrier.

Until quite recently the aim of one of the required tests was to kill an animal rather than trying to cure it. In this experiment, more and more of the drug is given to the animal of choice until it is poisoned and dies. The idea is to establish the difference between the dose at which a drug is fatally toxic and the dose at which it is effective in producing its desired therapeutic endpoint. If these doses are too close together, it would be relatively easy for somebody to inadvertently overdose, and the drug would not be safe to use. Problems like this have plagued some important drugs such as barbiturates, the first widely used sedative drugs. The death of Marilyn Monroe in 1962 from a barbiturate overdose was a dramatic public reminder of this kind of problem and prompted the government to require increased testing of new drugs.[16] In summary, passing the Food, Drug, and Cosmetic Act as well as its many subsequent iterations put the responsibility for drug safety and effectiveness squarely on the shoulders of the pharmaceutical companies. This transformed the science of drug development and necessitated the use of millions of animals every year for performing toxicity and other tests to ensure that pharmaceutical companies were in compliance with laws designed to ensure the safety and effectiveness of their products.

We can therefore see why there is such a demand for using animals in today's biomedical research enterprise. Animals are used throughout the world in university and hospital laboratories as well as in the pharmaceutical industry. As we have noted, the vast majority of experiments performed on animals are not designed to find out things about the animals themselves but are anthropocentric and are designed to find out things about humans in an effort to help them therapeutically. Except for a few regrettable examples, such as the work of Herophilus and Erasistratus in the second century CE (see Chapter 2), the "scientific" work of the Nazis and the Japanese during the Second World War, and some of the secret activities of the CIA, such as their MK-ULTRA program in the 1960s, we do not conduct experiments on humans unless they are very carefully monitored by specially constituted oversight committees in institutions like universities to protect human subjects from harm (see Chapter 8). This is the case, for example, when we conduct highly controlled clinical trials required by bodies like the FDA to determine if a new therapy really works in people the way a pharmaceutical company says it does. In other cases, animals are used as stand-ins for humans.

This raises several questions. Do the animal tests carried out by the pharmaceutical industry really tell you anything important about the effects of drugs on humans, and if so, what kinds of translational information can they provide? Another type of question that might interest us is, even if it is true that animal testing can tell us important things about humans, how often in practice does this really happen? Finally, are there any alternatives to the use of animals these days for safety testing by the pharmaceutical industry?

The Nature of Animal Models

There is a lot of literature available discussing what may be good or bad, useful or wasteful about animal research. Clearly, there are many people who feel very strongly about animals but because of their lack of training know little about science. Many people have a kneejerk reaction about experiments performed on animals and condemn them universally, without really knowing much about what they are designed to tell us and what scientists do.

We have already discussed several important examples of animal experimentation drawn from history, all of which were extremely cruel but clearly provided information that was of fundamental importance for our understanding of human biology. Consider all the experimental vivisections performed by Harvey, Galvani, Magendie, and Bernard. However repulsive their work may have been from a moral and ethical viewpoint, from a purely scientific perspective much of it would be considered excellent research. People like Claude Bernard are held in the highest regard in the pantheon of scientific greats.

But do all the animals that are used in basic and applied research these days really provide information that is even remotely as important as that provided by Harvey or Bernard? Let's consider animal models further—what might they be able to tell us that is useful, and what are their limitations? Let's also consider animal models purely from the point of view of scientific effectiveness, a question distinct from whether using such models is ethical or not, which we will discuss in Chapter 8.

To begin with, if you were to visit a large research building in a major university and walk around the laboratories, what kinds of animal models would you find most scientists using? Would you find them using a duck-billed platypus? No. Would you find them using a giraffe? No. In fact, there are relatively few animals that scientists use in their research on a quotidian basis. One of the reasons for this is that there is a huge corpus of knowledge that has accumulated around the use of certain animal models that scientists can avail themselves of and contribute to with their own results in a way that other scientists can readily comprehend. Conditions concerning the breeding, living conditions, and other features of these animals' lives are extremely artificial and standardized throughout the world so that ideally what goes on in one laboratory can be directly compared with what goes on in another laboratory. You don't just go to the pet store and pick up a few mice, take them back to your laboratory, and do some experiments. The mice you use are very particular "inbred" strains that are sold by specific vendors that everybody else uses. The conditions that prevail in the life of a laboratory mouse have nothing whatsoever to do with the way these animals would normally live in their natural environment. And that, of course, is precisely the point. Everything is standardized so that no untoward "variables" creep into experiments, except for things that the scientist specifically wants to study. If you want to study mice in their natural environment, then you would be studying natural history, not performing experimental biomedical research.

Highly standardized animal models are a central part of the *lingua franca* of scientific discourse. When a research scientist says the word "mouse," they don't just envisage a small furry rodent; in fact, they probably don't envisage a small furry rodent at all. Rather, they envisage a set of genes, a set of physiological systems, and a set of behaviors that are defined in their mind by the word "mouse." Nobody outside the world of science thinks about mice this way. They don't even have the technical language that would allow them to think this way. For most people a mouse is a just a mouse; it isn't a scientific construct or a model for anything else.

Which animals, then, are generally used by scientists as models for humans?[17]

As can be seen from Table 4.1, the list runs the gamut from animals that are usually considered to be very much like humans, such as primates, to things like flies and worms, which aren't. In addition to this general list, there are some animals that are frequently used for specific purposes, such as in the pharmaceutical industry, where pigs, dogs, and rabbits are commonly used in ADMET studies. If you're not a biologist, it might seem absurd at first glance that worms and flies are used as "models" of humans, but as we shall see, in some instances they really can be useful. As Nietzsche wrote in *Also Sprach Zarathustra*, "Ye have made your way from the worm to man, and much within you is still worm."

Of the vertebrate animals on the list, rodents do most of the heavy lifting these days, and frankly, if you are a mouse, you are in good shape. The fact is that every human disease has been cured umpteen times when it is modeled in mice. As the renowned cancer researcher Dr. Judah Folkman said, "If you have cancer and you are a mouse, we can take good care of you." The scientific literature is replete with reports that mice who have strokes, cancer, pain, bubonic plague, clinical depression, or any other disease known to humanity have been completely "cured" by some fantastic new therapy. Unfortunately, most of this is just nonsense. People still suffer from all these things, so clearly something must be wrong with these mouse models of human disease. But,

Table 4.1 Model Organisms and Their Common Names

Model Organism	Common Name
Saccharomyces cerevisiae	Yeast
Drosophila melanogaster	Fruit fly
Caenorhabditis elegans	Roundworm (nematode)
Danio rerio	Zebra fish
Mus musculus	House mouse
Rattus norvegicus	Common rat
Macaca mulatta	Rhesus macaque

wrapped up in their own jargon that nobody else can understand, scientists today are often guilty of peddling the same kinds of fantasy cures that the purveyors of snake oil did in the nineteenth century.

Something that is not usually discussed but will be obvious is that most of the main animals on the list like rats, mice, worms, and flies are things that humans generally have an aversion to. Cat, dog, and monkey use has decreased greatly in Western Europe and the United States since Magendie and Bernard's time. The animals used nowadays most often have traditionally been thought of as "vermin," purveyors of death and disease. Naturally, this allows people to more readily accept anything that is done to them, and scientists find it easier to talk about what they are doing to them in the public arena. People generally don't like vermin, which engender feelings of disgust in them, as suggested by the Nazis' frequent references to the Jewish people as *Ungeziefer* (vermin). As far as most people are concerned, the effect would have been quite different if they had opened a book and read that "Gregor Samsa awoke one morning after a night of disturbing dreams to find that he had been transformed in his bed into a giant bunny rabbit." However, it is a strange irony that the mice that researchers use nowadays are more germ free, cleaner, and less likely to spread disease than the researchers who experiment on them.

What kinds of animals would be the best models for humans, and under what circumstances? What kinds of things can we potentially learn about humans from studying other animals? Well, to begin with, I am sure that we would all have to agree with Alexander Pope, who said, "The proper study of mankind is man." Anytime we study an organism that isn't a human, something must be different—something must be lost. Even if we took Claude-Nicolas Le Cat's idea and built a machine to mimic a human, as pointed out by Borges at the beginning of this chapter, a machine that replicated all the facets of a human would obviously be another human. Maybe one way of starting to think about all of this is to think about genes, because in many respects it is our genes that fundamentally direct what kinds of animals we turn out to be. Also, genetics is a "hot" topic these days, and the media constantly bombard the public with genetic information as a touchstone when describing humanity and its similarity to other species. We are told, for example, that human beings are about 99% similar to one another in terms of their genetic makeup. The small differences that exist are responsible for all human diversity, including the propensity to develop different kinds of diseases.

But if we can't use humans for our experiments, let us consider what animals are the most like humans genetically. The answer, perhaps not surprisingly, is chimpanzees and bonobos, which appear to be around 98% genetically like humans. But what does this mean? When you see a chimpanzee, how similar do you think it is to a human? There certainly seem to be many similarities and differences. We should note, by the way, that humans aren't automatically "better" than chimps. Although humans may be superior to chimpanzees when it comes to things like language, chimpanzees are better at other things, like climbing trees. Would 98% similarity seem like a reasonable number? In any event, chimpanzees do really seem rather like humans in numerous respects, so perhaps they would make good animal models for many purposes.

What about a mouse? Mice have become the most widely used stand-in for humans for a variety of reasons.[18] They are certainly cheaper and easier to breed and use than monkeys. Almost every gene found in one species so far has been found to exist in the same or a closely related form in the other. Let's run the numbers. Both the mouse and human genomes contain about 3.2 billion base pairs, the chemical letters that constitute genetic information. Interestingly, however, only between 1% and 2% of the human or mouse genome consists of protein-coding regions (genes). Most of the rest of it is made up of noncoding DNA, sometimes called "junk" DNA, that has a somewhat mysterious function. It is 1% to 2% of the genome that directs the synthesis of proteins, and this is the part we are most interested in when people talk about "similarities." On average, the protein-coding regions of the mouse and human genomes are 85% identical. Some genes are as much as 99% identical, while others are only 60% identical, so this figure is an average. When you look at a mouse, how similar do you think it is to a human? If chimpanzees are 98% the same, does 85% similar seem correct for a mouse?

How about a banana? Data in the literature says that humans and bananas are around 60% similar genetically.[19] Does that seem reasonable to you? More than likely, you think that you have a much better memory than a banana, that you are more articulate, that you can run faster, and all kinds of things. On the other hand, a banana is much better than you at being yellow. I would bet that you think that a banana really isn't 60% like you. You would most likely say that it is an overestimate and that nobody you know is even remotely like a banana, certainly not 60% similar. Nevertheless, however we look at it, it is clearly true that there are quite a few genetic similarities that we share even with things as simple as a banana or even a yeast cell, one of the most basic eukaryotic (i.e., nucleus-containing) cells whose genome is around 25% similar to that of a human. If that is the case, then perhaps there are certain things we could learn even from yeast or bananas that would be helpful for understanding humans, although we would surely agree that the *elan vital* of a yeast cell, a banana, or a human is totally different.

Of course, one of the great mysteries in genetics today is to ascertain what exactly most of the genome, that is, the greater than 90% of it that isn't coding for proteins, is for. Here the similarities in sequences drop a great deal when we consider different human individuals or different types of animals. One important idea is that there are many sequences in these regions that can act to regulate the activity of the protein-producing genes. A huge project called ENCODE seeks to answer such questions by fully describing these *cis*-regulatory elements (CREs). Recently, almost an entire issue of the journal *Nature* was devoted to the latest results of this project. It is clearly true that the differences between individuals and species are due not only to the actual proteins they make but also to how this entire process is regulated.[20]

What kinds of things, then, might we learn about humans from yeast cells or other primitive organisms? A good way to approach this question is to think about how animals have evolved. One can imagine, as Darwin's follower Ernst Haeckel did in the nineteenth century (see Chapter 6), that the evolution of all living things is like a giant "phylogenetic tree" with many branches.[21] The pattern of branching of the phylogenetic tree reflects how species evolved from a series of common ancestors, and when looking at such a tree, two species are considered to be more related if they have a more recent common ancestor and less related if they have a less recent common ancestor. For example, the lineages leading to modern rodents and primates are thought to have branched off from one another or "diverged" from a common ancestral species that lived some eighty million years ago.

Some organisms that exist lower down in the original trunk or roots of the tree represent our most ancient living ancestors and will have things in common with just about every other organism. What kinds of things might these be? We know that all forms of life, even things as ancient as bacteria, have genes that need to be duplicated, transcribed, and translated into proteins. Processes like this even occur in viruses that are only "sort of" alive—they are really parasites that can only exist with the help of cells. We can see, therefore, that if we study gene regulation in bacteria or viruses, we may learn things that are still true in humans because the basic principles of the process have remained the same throughout evolution. Genes in both humans and viruses are made of strings of DNA or RNA. In fact, recent research has dramatically illustrated the incredible importance of some of these results. More than twenty years ago, biologists Andrew Fire and Craig Mello uncovered a process called RNA interference in the lowly worm *Caenorhabditis elegans*—a groundbreaking discovery that earned the pair the 2006 Nobel Prize in Physiology or Medicine. Just a few years ago, the Nobel Prize was awarded to two researchers who discovered the gene editing tool CRISPR/Cas9 when they were performing research on how bacteria resist infection by viruses. Now these two methods have been applied to humans to provide an extremely promising approach for the treatment of sickle cell disease,[22] the first real use of human gene editing approaches in human therapeutics. This is a fantastic example of how basic research based on analogies between humans and other creatures can be of direct benefit to humanity. Naturally, there have been many changes in the details of exactly how gene regulation works, which have developed over the long course of evolution, but just as Harvey's work on animals led to his idea of the circulation of blood in humans (see Chapter 3), we may get to understand some of the central and most basic concepts about how human genes are replicated from studying an organism as lowly as a worm, a bacterium, or a virus. In fact, some of the most important scientific information that originally showed that genes were made of nucleic acids (DNA and RNA) came from studies using bacteria and viruses.

On the other hand, what if we wanted to have an animal model of consciousness, something that many people think is unique to humans and clearly plays a vital role in the symptomology of many human diseases— would we think that we could study such a thing in bacteria? Definitely not! A banana? Very unlikely. A mouse? Now we are getting into a grey area. Is there any way that studying this animal with its 85% genetic similarity with humans could be used to study consciousness? For that to be possible, we would have to have a clear idea of what consciousness is in material terms, and then see if mice had it as well. This is something we will discuss in Chapter 5.

Another important reason animals such as fruit flies and worms are used as models for humans is that their genes can be easily manipulated to produce changes called mutations. When a gene is mutated to produce a slightly different nucleic acid sequence, we can observe the consequences of these changes for the biology of the cell or animal concerned, and this may give us a clue as to what the gene normally does and how it does it. Furthermore, mutations in genes that have adverse consequences are often the prime causes of human diseases.

Let's consider the cardiovascular system, which has been the touchstone for our discussion of scientific experimentation. In fact, some of the genes that control blood flow in a fruit fly (fly blood is called hemolymph) are the same genes that contribute to the human cardiovascular system. Of course, the human system looks very different from that of a fruit fly. Nevertheless, the fly does have evolutionary ancient bits and pieces that are analogous to the human cardiovascular system, and their formation is determined by some of the same genes as in a human.[23] The "seeds" of the human system are indeed contained in that of the fly. Strange as it might seem, mutating genes in fruit flies and analyzing changes that occur in the fly cardiovascular system can provide us with ideas about how the same things might happen in a human, particularly when it comes to questions like which genes are involved in the embryological development of the system overall, something that could be extremely helpful if we want to understand heart diseases in developing embryos or young children.

But what if we want to go beyond significant developmental genes to actual physiology, the way organs function in a mature human? Here we may well need an animal model that is closer to a human. Originally, the cardiovascular system arose during evolution to solve a problem—how to get oxygen distributed throughout multicellular bodies that were too large to use simple diffusion of the gas to do the job.[24] The energy that is released from reactive oxygen is what powers life in living objects. In Sir Thomas Browne's words, "Life is a pure flame, and we live by an invisible Sun within us."[25] Rather than releasing the energy of this chemical reaction all at once, which would be dangerous, like the burning of a flame, cells deconstruct oxygen's chemical reactivity step by step using a series of biochemical reactions so that the energy released can be distributed in a manner commensurate with living systems. The origin of multicellular animals (metazoans) dates back approximately 770 to 850 million years. The most primitive living phyla of multicellular animals are the *Porifera* (sponges), followed by *Cnidaria* (corals and jellyfish). Between 600 and 700 million years ago, a new body plan evolved that exhibited bilateral symmetry. The precursors of the human cardiovascular system in animals arose very early in evolution and in many different ways. Some animals evolved a closed circulatory system like humans, whereas other animals evolved an open system of the type originally envisaged by Erasistratus and Galen. Furthermore, the movement of blood is powered by a heart-like pump in some animals and by the pulsating movement of arteries in others, something that Galen also imagined as a possibility (see Chapter 2).

In an animal, as in a human, the cardiovascular system is used for many purposes, such as the transport of substances (e.g., nutrients, oxygen, carbon dioxide), hydraulic force generation (e.g., head-foot protrusion in mollusks and penile erection in vertebrates), regulation of heat (e.g., via countercurrent flow), ultrafiltration (e.g., in the kidney), defense (e.g., through delivery of clotting factors, immune factors/cells), and whole-body integration (e.g., hormonal regulation). So how would studying other animals help us understand human cardiovascular physiology? Let us return to William Harvey's experiments. Harvey made use of an enormous number of animals during the twenty years of experimentation he carried out attempting to understand how the human cardiovascular system worked. When using warm-blooded animals like dogs, he found that the rapid contraction and relaxation of the heart made it difficult for him to make accurate observations. So, as we have seen, he had a good idea. He examined cold-blooded animals like reptiles and found that the rate of the heartbeat was much slower, making it easier to see it going through its normal cycle of contraction and relaxation. Harvey also found it convenient to perform experiments on reptiles in which he applied tourniquets to veins and arteries to see how this affected blood flow and observed on which side of the tourniquet blood built up, enabling him to work out the direction of blood flow.

Now let us consider how good a snake is as a model for the human cardiovascular system. Clearly, Harvey used snakes to his advantage. However, his experiments would only have been informative if those aspects of the snake cardiovascular system he was studying really resembled the human situation. And fortunately for Harvey, they did in many important respects. First of all, snakes do have a closed circulating cardiovascular system like humans. The path that the blood takes, including its transport through the lungs, is the same. Hence, if what you are trying to establish is the fact that, as opposed to Galen's idea of an open system, animals can have closed cardiovascular systems, and that this is the way blood circulates around the body, then the snake is a reasonable model.[26]

But if you wanted to go further in your analysis of the human cardiovascular system and find a good model for human cardiovascular diseases, you would find that there are plenty of things about the way a snake's cardiovascular system works that are not similar to humans. For example, the snake heart only has three chambers, rather than the four that exist in humans or other mammals. Because of the three-chambered arrangement, many important details concerning exactly how blood flows through the reptilian heart are quite different from mammals and from humans in particular. A mouse might be a better model for a human because it is closer to a human phylogenetically. A mouse has a four-chambered heart, and some important details of mouse cardiovascular physiology are closer to a human than those of a snake. Hence, when answering certain questions about human biology, a mouse might be appropriate and better than a snake. But the 15% lack of identity between humans and mice genetically still gives us a lot of scope for functional differences, and this is why one cannot be surprised that in many respects, mice are far from ideal stand-ins for humans. Nevertheless, we can conclude that it is certainly possible to find out important things about humans from studying animals or even organisms that are extremely simple.

Clearly, the utility of using animals for research will be critically dependent on exactly what questions we are trying to answer—and this cannot be stressed too often. Sometimes an animal like a mouse is a good model for a human, and sometimes it isn't. When a scientist publishes a paper using mice as a model for humans, a critical assessment of precisely what can be translated to humans (assuming that is the goal) is essential; but unfortunately, this is very rarely done very effectively. And this leads us back to the situation mentioned above: every human disease has been "cured" in mice multiple times over.

Sometimes the lack of analogy between animals and humans can also have important practical consequences. As we have seen, one of the situations where millions of animals are used every year is in the pharmaceutical industry for ADMET studies. Such studies are supposed to indicate that a drug is safe enough to be moved into human clinical trials and provide some preliminary information about how it is metabolized and appropriate human dosing. But how useful are animal models for these translational goals? As we discussed above, the laws requiring this kind of research were first introduced following the sulfonamide elixir debacle of 1937. Today, the FDA generally requires preclinical testing of any new drug or biological therapeutic "for pharmacologic activity and acute toxicity in animals." In principle, this certainly seems like a good idea, and one can see why it was initially instigated. The problem with these studies is that, in general, they don't work very well.[27] What do I mean by this? Basically, there are two types of things we want to know if we are developing a drug for use in humans: Does it produce the effect that we had in mind, and if it does, are there other factors that might make it dangerous or impractical for humans to use? According to Hodgson:[28] "A chemical cannot be a drug, no matter how active nor how specific its action, unless it is also taken appropriately into the body (absorption), distributed to the right parts of the body, metabolized in a way that does not instantly remove its activity, and eliminated in a suitable manner—a drug must get in, move about, hang around, and then get out."

As it turns out, the way a drug is metabolized can have extremely important consequences for the way in which it works. Some drugs are completely inactive if they aren't metabolized properly by animals. Consider heroin, for example. Heroin, as we have described, is diacetyl morphine. Opioid drugs work in the body by binding to and activating a receptor protein called the μ-opioid receptor. Morphine activates this receptor, which is situated at points in the nervous system where it can control pain. Heroin is inactive at this receptor. It doesn't activate it at all. However, when heroin is given to most animals, enzymes remove the two acetyl groups, regenerating morphine, which is then active.[29] In other words, heroin is a "pro-drug," an inactive precursor molecule whose action in an animal is absolutely dependent on its correct metabolism. It is heroin's metabolite morphine that is responsible for its effects.[30]

Drug toxicology and metabolism studies that are carried out by the pharmaceutical industry are extremely complex, and this is not the place to go over them all in detail. However, it is rather ridiculous to imagine, even in principle, that toxicological studies using animals would reliably predict adverse effects in humans. Let us say you design a drug that is supposed to act on a particular target in a human—a β-blocker for cardiovascular disease, for example. The target of this drug would be the β-adrenergic receptor, which is very similar in a mouse and a human. To be precise, it is 93% similar at the genetic level, and drugs that work on mouse receptors work on human receptors in a rather predictable way. If the drug works in a mouse, we can be pretty sure it will work in a human. But when testing for possible adverse side effects in general, there are plenty of detailed differences between animals and humans in things like the enzymatic pathways that control drug metabolism, the way a drug is absorbed, the way it is secreted, and innumerable unknown and therefore unanticipated "off target" effects that may be problematic.

Why would we imagine that, when all these things are considered together, the results in a mouse or another animal would reflect the overall situation in a human? It is true that certain dangerous drug side effects turn up often enough that they can be specifically tested for in advance. But there are innumerable others that one could not anticipate—the human "toxome" and that of a mouse is substantially different. And the results speak for themselves: an analysis of 2,366 drugs concluded that "results from tests on animals (specifically rat, mouse and rabbit models) are highly inconsistent predictors of toxic responses in humans and are little better than what would result merely by chance—or tossing a coin—in providing a basis to decide whether a compound should proceed to testing in humans."[31] Yes, it is certainly true that animal toxicological tests do sometimes predict problems that occur when a drug is subsequently given to humans, but frankly, tossing a coin is a much cheaper way of going about things, and is certainly much less cruel to animals.

There are numerous examples of how the lack of predictive value of animal trials in the pharmaceutical industry has led to disastrous problems when drugs are given to humans. I will just discuss one recent example for illustrative purposes. An area of biology that is of great current interest concerns the pharmacology of cannabis, which has come to prominence once again after some seventy years in the scientific wilderness. Thousands of years of human experimentation have suggested that cannabis might be useful for treating a number of human problems including pain, nausea, anxiety, and many other things. The effects of cannabis are mostly due to its major pharmacologically active component, the chemical delta-9-tetrahydrocannabinol (THC). However, the effects of THC are reflective of the fact that all animals have their own endogenous cannabis-like system. This involves chemicals called endocannabinoids that act on tissues that express cannabinoid receptors. THC works because it mimics the actions of these substances and activates the same receptors. Animals have enzymes that synthesize endocannabinoids and metabolize them. Thinking about this arrangement allows us to come up with an interesting drug strategy. If we make a drug that inhibits a major enzyme that normally destroys endocannabinoids (the enzyme is called fatty acid acyl-hydrolase, or FAAH), then our endocannabinoids will not be destroyed as normal and their levels in our bodies will increase accordingly. Increased levels of endocannabinoids should produce more activation of cannabinoid receptors and produce effects that are similar to those produced by THC, many of which might be predicted to be pharmacologically useful. Once this fact was realized, the race was on to produce a drug of this type and several pharmaceutical companies produced a range of potential FAAH inhibitors.

One company that was in the game early on was a Portuguese drug company called Bial. The company began a series of preclinical trials on new FAAH inhibitors it had made and picked a drug called BIA 10-2474 as its lead molecule for further development. After conducting a series of animal experiments and ADMET tests, the company decided that the drug seemed promising and safe enough to begin clinical trials in humans. Human clinical trials usually begin with Phase 1 trials that are run to assess the general safety of the new compound and to collect some basic data on drug metabolism. To carry out this trial, Bial hired an experienced company in Rennes, France, named Biotrial Research that uses hospital facilities at Rennes University for carrying out trials such as these. Six doses of BIA 10-2474 were to be tested on humans. Each dose was to be administered for ten consecutive days. Nothing untoward was expected, yet things didn't go according to plan. In fact, the trial was a complete disaster.[32]

On the evening of day five (January 10, 2016), one of the six volunteers who had received the drug rather than the placebo had to be hospitalized in serious condition. This patient went into a coma and died. Four of the other five men in the same dosage group were also hospitalized between January 10 and 13 displaying similar symptoms to those of the man who died. The symptoms experienced by the affected participants were all neurological in nature, resembling those associated with a stroke. These included severe headaches and problems with consciousness, ranging from sedation to coma and memory impairment. The trial was immediately abandoned. The four remaining hospitalized patients all eventually survived, although several of them had severe symptoms over an extended period of time. Brain imaging studies demonstrated that these individuals had several structural anomalies in their brains strongly suggestive of a toxic or metabolic mechanism for the adverse effects of the drug.

Subsequent investigations as to the causes of the fiasco revealed several interesting things. As it turned out, BIA 10-2474 wasn't a very good FAAH inhibitor in humans. The potency of the drug wasn't very high, and its selectivity, meaning its propensity to act on FAAH rather than other targets, also wasn't very clear. It appears that the animal testing done by Bial with respect to effects on pain, which had been the major endpoint to be examined, was simply not predictive of the problematic effects seen in humans, and the doses calculated for use in humans were based on animal experiments, which turned out to be completely misleading. A key question was still whether the toxicity of the drug was due to its effects on FAAH or due to other "off target" effects resulting from its lack of specificity. We now know that the drug had numerous effects on proteins in the body in addition to FAAH and that these effects were specific to humans and not picked up at all by the animal studies. Hence, it is clear that in the case of this dramatic example, there was a complete failure of animal and ADMET testing to predict the awful consequences of giving this drug to humans.

To be fair, the nonsensical nature of many of the toxicology trials conducted by the pharmaceutical industry has, by this time, become apparent to the administrative bodies that are concerned with coming up with and enforcing drug trial regulations, and there are ongoing attempts to move toward abandoning some of them in favor of other non-animal- based tests (see Chapter 6). The current laws are constantly being revised in order to get companies to use human-based tests in ADMET and drug evaluation studies. Nevertheless, we are far from ridding ourselves of this general approach; drug companies are only obeying the law, after all, and the laws that cover these things are woefully out of date. Fortunately, there are increasing efforts nowadays to develop non-animal-based methods for predicting how drugs will work in humans, using systems like human organoids (see Chapter 6). Indeed, the first new drug has recently been presented to the FDA for approval based on testing that was completely animal free. We will return to this topic when we discuss current alternatives to animal testing (see Chapter 6).

The various problems with the millions of ADMET tests that are carried out by the pharmaceutical industry on unfortunate animals every year certainly highlight what happens when animals are used as inappropriate models of humans. This brings us back to animals like rats and mice, which serve as the main animal models used as stand-ins for humans in basic biomedical research nowadays, not only in the pharmaceutical industry, but also in thousands of universities, hospitals, and research institutes throughout the world. We have seen that, depending on the question asked, a mouse might furnish useful information about some important basic aspect of human biology. But more often than not, experiments carried out by scientists in places like universities do not do this. As in the pharmaceutical industry, the results obtained by scientists in universities rarely translate to produce anything therapeutically useful in humans and yet hundreds of millions of mice and other animals are used for this purpose every year. So why?

The Use of Animal Models

It is clear that in the nineteenth century, Magendie and Bernard's vivisection experiments really represented the apogee of cruelty to animals carried out in the name of science. Just cutting up a live animal without using anesthetics or pain-killing drugs is the stuff of nightmares. As that century wore on, things started to improve somewhat. First, there was the discovery and use of ether, chloroform, and later intravenous barbiturates and analgesics to produce surgical anesthesia, analgesia, and amnesia. Second, there were advances in antiseptic techniques and the use of antibiotics that helped reduce infections when animals underwent surgery.

Although these types of interventions certainly reduce some of the acute agonies experienced by animals at the hands of scientists, they allow for the introduction of survival surgery, that is, the continued use of an animal following an experimental procedure from which it has recovered. Nowadays, increasingly sophisticated surgically prepared animal models that are supposed to mimic human diseases are constantly being produced to serve current research needs. These animal surgeries result in the preparation of conscious animals surgically instrumented with catheters, electrodes, and telemetry devices, which have become the mainstays of basic animal research in some fields. In the experiments described in Chapter 1, I discussed just this type of thing. The experiments I described used survival surgery to produce a mouse model of disease in which the resulting animal was well endowed with numerous highly invasive recording devices that would allow the experimenter to obtain data describing neuronal activity and behavior in real time. Experiments like these are nearly always performed on mice. There are numerous reasons for this. One of the most important is that we can now do a different kind of surgery on mice; we can do "surgery" on their genes. Our knowledge of the structure and function of the mouse genome allows scientists to use molecular biological methods to make changes in different mouse genes and observe the results. From this point of view, the mouse has become the mammalian equivalent of the fruit fly, the traditional animal for examining the effects of genetic mutations.[33]

Let us say that a human disease has been observed to be associated with a mutation in a particular gene. A scientist will typically think that this suggests an interesting experimental possibility. Use the science of molecular genetics to make a line of mice that expresses exactly the same gene mutation as found in the human disease and, *voila*, one has created a mouse model of the disease—or at least that is the assumption. Unfortunately, things aren't that simple, and the gene mutation expressed in a mouse often doesn't mimic the human disease or has completely different consequences to those observed in a human. Scientists have a tendency to think that if they are using a technique that involves employing the latest technical advances, then they are automatically performing more meaningful experiments and providing better answers to questions. Regrettably, nothing could be further from the "truth."

Mice are currently used as models for virtually every aspect of human biology and disease. Indeed, mice represent by far the largest percentage of animals used in biomedical research (greater than 95%), so it is particularly important to evaluate how useful they are for this purpose. How similar is a mouse to a human from a functional point of view? Well, as we have seen, mice and humans are genetically around 85% similar, which is an interesting number, but what does it mean in practice? We have seen that in the pharmaceutical industry, differences between mice and humans regularly lead to animal data not being reliably translated to humans. How about in the realm of basic research? A mouse, like a human, has a bilateral body plan. It has four limbs. It has organs that are similar to those of a human, including a cardiovascular system, a brain, a liver, kidneys, and so on. Mouse reproduction and development take place in a similar way to a human. Mouse cells mostly function like human cells. But, as we have pointed out, a mouse isn't just a small furry human being. And indeed, current research using genetics proves the point, particularly if we begin to dig deeper than the 85% similarity number.

A recent lead article in the journal *Nature* compared genetic markers in the cerebral cortex of the brains of mice and humans.[34] The authors observed that "Comparison of similar mouse cortex single-cell RNA-sequencing datasets revealed a surprisingly well-conserved cellular architecture that enables matching of homologous types and predictions of properties of human cell types. Despite this general conservation, we also found extensive differences between homologous human and mouse cell types, including marked alterations in proportions, laminar distributions, gene expression and morphology. These species-specific features emphasize the importance of directly studying human brain." They concluded, "Furthermore, these results help to resolve the paradox of failures in the use of mice for preclinical studies despite conserved structure across mammals and highlight the need to analyze human brains in addition to model organisms." Reports like these are common these days when comparing mouse and human tissues. In other words, even overall system similarities may hide important molecular and functional differences when comparing the two species, something that is almost invariably true.

Nevertheless, in some cases, as we have seen, a mouse can act quite adequately as a stand-in for a human. Consider Paul Ehrlich's work on the development of Salvarsan. Here he used thousands of animals such as mice and rabbits that he infected with *T. pallidum*. This organism is infectious in rodents, so, issues like differences in metabolism aside, giving them newly created drug-like molecules might allow one to see if they are effective in clearing the infection. Basically, Ehrlich and his colleagues were using animals as living Petri dishes.

However, it isn't always that easy to model infectious diseases in animals. Consider the HIV-1 virus, something that played a similar role in the twentieth century to *T. pallidum* in the nineteenth. In this case, the "tropism" of the virus results in it only being able to infect human cells. There are similar viruses such as SIV that infects monkeys and FIV that infects cats, but HIV-1 is completely specific to humans. This fact hasn't prevented thousands of scientists from trying to devise mouse models of HIV-1 infection. HIV-1 infectivity in humans involves several steps, including binding and entry of the virus into its target cells, viral reproduction, and so on. Some mouse models have been able to introduce one of these steps into mice, but trying to successfully model the entire viral life cycle and its effects on the immune system and brain has not been possible and probably never will be in any useful manner. This means that there are no mouse models that can be used for developing therapies for AIDS, including things like HIV-1-related dementias. Nevertheless, every year the government gives out hundreds of millions of dollars in research funds to support animal-based projects that are unlikely to ever produce anything useful in terms of treating human HIV-1 dementia. How many useful new drugs have come from the use of current mouse models of HIV-1 dementia? None. There are many other types of animal models that generally fail to adequately represent human diseases. Cancer is another notorious problem as evidenced by the high clinical failure rate of cancer drugs developed through animal experimentation, among the highest for any disease category.[35]

When thinking about many aspects of physiology and disease, one must consider a human as an entire system of interacting components, and here we should give Claude Bernard, whatever his faults, some credit for his ideas framing animals as homeostatic systems. An animal is a set of dynamic moving parts always interacting with one another in an attempt to achieve a viable physiological state. There are many components that interact when determining the ultimate set of symptoms a human or an animal patient will display. Let us consider the cardiovascular system once more. One of the most common and deadliest diseases to afflict humans is a stroke. In an ischemic stroke (the most common kind), a vessel that supplies blood to the brain becomes occluded, so the brain doesn't receive an adequate blood supply. The normal physiology of the brain requires a constant source of oxygen supplied by the blood, and even a brief interruption of this process can lead to serious brain damage. The precise molecular mechanisms that cause this brain damage are now understood to some degree, but efforts to produce meaningful models of a stroke in mice have completely failed. Important details of a mouse's blood circulation are not the same as those of a human, and mice do not normally develop strokes. Yes, it is possible to occlude blood flow to a mouse's brain using a number of artificial methods, and these procedures result in brain damage that resembles a human stroke in some respects. However, the situation in a human is much more complicated. Many stroke patients also have additional factors that affect their health and contribute to their propensity for having a stroke. These include hypertension, diabetes, and smoking; there are many others. Just making a hole in a mouse's brain really doesn't get you very far without considering things like this. The results are all, frankly, useless. Like the HIV-1 project, millions of dollars have been spent on research where scientists perform procedures in which they destroy the brains of mice. Again, we may ask, what has come of all of this, and how many useful drugs or other forms of therapy for treating stroke patients have resulted from these mouse studies? The answer is, once again, none. Nevertheless, there are gazillions of scientific publications every year claiming to have produced helpful interventions for stroke based on mouse experiments.

It will be clear that some things are more difficult to model in mice than others. Human diseases that affect the higher functions of the nervous system are particularly prone to these kinds of problems. How do we model human conscious behavior and cognitive abilities using a mouse? What exactly is the consciousness of a mouse like (see Chapter 5)? Consider something like traumatic brain injury, a common problem. Here again, we could use a mouse model. Scientists just drop a weight on a mouse's head. Has anything useful come of this? Nothing. How about spinal cord injury? It is easy to mimic that in a mouse. Just drop a weight on its back, break it, and spend hundreds of millions of dollars examining what happened. And what have we humans gained from all of this? Again, nothing. Then there are chronic degenerative diseases of the nervous system like Parkinson's disease, Huntington's disease, and Alzheimer's disease. Have there been any useful results that have produced new treatments for these diseases? Very few.[36] Overall, the uncritical use of animal models together with other phenomena such as the "crisis of reproducibility" (see Chapter 9) has resulted in an enormous number of animals being routinely killed for little gain.

Probably the most common way to find new therapeutic uses for drugs is for doctors to experiment with existing drugs "off label" and repurpose them. In his recent book *Pharma*, Gerald Posner puts it this way:[37]

> The research divisions of pharmaceutical companies were always on the lookout for additional therapeutic uses for their existing product line. Surprisingly few, however, were discovered in the laboratory. Instead, most of the breakthroughs were the result of the unregulated power of doctors to dispense a company's drugs off-label. Physicians are presumed under the law to have enough expertise to use drugs as they deem fit. They can change the approved dosing, prescribe one approved drug to be taken in combination with others, or use it for conditions unrelated to the clinical testing for safety and efficacy conducted by the drug company.

As science has progressed in recent decades, it has become more and more technical and, as a result, increasingly expensive. Many of the microscopes utilized in a laboratory routinely cost hundreds of thousands of dollars, and other pieces of equipment that are employed in university "cores" for use by many different investigators can cost even millions of dollars. In effect, producing virtually nothing in the way of useful translational data has become increasingly expensive—certainly not an encouraging trend, at least unless we change the way we do things. Interestingly, most pharmaceutical companies have realized what is going on and have abandoned research in some areas of neuroscience, neurodegenerative disease, and neuropharmacology, at least for the time being.

Scientists are supposed to think scientifically. They do experiments and then analyze the results and move forward accordingly. No emotions are allowed in this process. Only a dispassionate examination of the facts is permitted. We have done "the mouse experiment," and it didn't work very well. Then perhaps we should ask ourselves why is this the case, and how we can change our modus operandi? In the meantime, the cruelty to animals involved in performing so many experiments. which we are told by the government and the scientific establishment are absolutely necessary if we are going to make progress in helping humans, might be characterized as rather the opposite. The emperor, it seems, has no clothes. Isn't it time to stop doing this kind of thing and try something else? But what?

The use of animal models has been at the very heart of the entire scientific enterprise since its very beginning, and there is no doubt that it is an approach that can yield important results. However, it very much depends on the precise nature of the question being asked, and it isn't at all clear that it is generally used these days in a manner that bears much critical scrutiny. Clearly, if the aim of most of this research is anthropocentric, experiments using humans would be much more appropriate. In the past, this hasn't been possible, but as we shall see, new advances have made human-based experimention much more feasible to the point that animal experiments are no longer necessary (see Chapter 6).

Moreover, aside from scientific arguments, we are yet to consider the ethical aspects of all of this. Whatever aspect of the problem we are considering, the use of animals as experimental models raises the question as to the conscious experiences, if any, of Descartes's animal machines, a subject to which we will now turn our attention.

Notes

1. Jorge Luis Borges and Andrew Hurley, *Collected Fictions*, Penguin Classics Deluxe Edition (New York: Penguin Books, 1998).
2. Jessica Riskin, *The Restless Clock: A History of the Centuries-Long Argument over What Makes Living Things Tick* (Chicago: University of Chicago Press, 2016), 137–138.
3. David Grimm, "Controversial Study Says U.S. Labs Use 111 Million Mice, Rats," *Science* 371, no. 6527 (January 2021): 332–333, doi:10.1126/science.371.6527.332; Larry Carbone, "Estimating Mouse and Rat Use in American Laboratories by Extrapolation from Animal Welfare Act-Regulated Species," *Scientific Reports* 11, no. 1 (December 2021): 493, doi:10.1038/s41598-020-79961-0.
4. Grimm, "Controversial Study Says U.S. Labs Use 111 Million Mice, Rats"; Carbone, "Estimating Mouse and Rat Use in American Laboratories by Extrapolation from Animal Welfare Act-Regulated Species."
5. Adrienne Mayor, *Poison King: The Life and Legend of Mithradates, Rome's Deadliest Enemy* (Princeton, NJ: Princeton University Press, 2011).

6. Simon Garfield, *Mauve: How One Man Invented a Color That Changed the World* (New York: Norton, 2002).

7. Bayer put their two new drugs, aspirin and heroin, on the market on exactly the same day in 1898; see Richard J. Miller, *Drugged: The Science and Culture behind Psychotropic Drugs* (Oxford University Press, 2015).

8. Paracelsus was a famous sixteenth-century alchemist who devised the iatrochemical method of treating individual diseases with specific chemicals, as opposed to the traditional approach of trying to rebalance different humors, which had been inherited from the work of Galen (Chapter 2).

9. Steven Riethmiller, "From Atoxyl to Salvarsan: Searching for the Magic Bullet," *Chemotherapy* 51, no. 5 (2005): 234–242, doi:10.1159/000087453; K. J. Williams, "The Introduction of 'Chemotherapy' Using Arsphenamine—the First Magic Bullet," *Journal of the Royal Society of Medicine* 102, no. 8 (August 2009): 343–348, doi:10.1258/jrsm.2009.09k036.

10. Kent A. Sepkowitz, "One Hundred Years of Salvarsan," *New England Journal of Medicine* 365, no. 4 (July 2011): 291–293, doi:10.1056/NEJMp1105345; Riethmiller, "From Atoxyl to Salvarsan"; Williams, "The Introduction of 'Chemotherapy' Using Arsphenamine."

11. Williams, "The Introduction of 'Chemotherapy' Using Arsphenamine"; Sepkowitz, "One Hundred Years of Salvarsan."

12. Anita Guerrini, *Experimenting with Humans and Animals: From Galen to Animal Rights*, Johns Hopkins Introductory Studies in the History of Science (Baltimore, MD: Johns Hopkins University Press, 2003), 114–136.

13. S. W. Gillian, "The Song of the Poison Squad," Group Website, Grace Wang & Abby Orler, The Pure Food and Drug Act of 1906, October 1903, https://42265766.weebly.com/song-of-the-poison-squad.html.

14. Carol Ballentine, "Sulfanilamide Disaster," *FDA Consumer Magazine*, June 1981, https://www.fda.gov/media/110479/download.

15. Arne Krüger et al., "ADME Profiling in Drug Discovery and a New Path Paved on Silica," in *Drug Discovery and Development—New Advances*, ed. Vishwanath Gaitonde, Partha Karmakar, and Ashit Trivedi (London: IntechOpen, 2020), doi:10.5772/intechopen.86174; Katya Tsaioun, "Evidence-Based Absorption, Distribution, Metabolism, Excretion (ADME) and Its Interplay with Alternative Toxicity Methods," *ALTEX* 33 (2016): 343–358, doi:10.14573/altex.1610101.

16. Miller, *Drugged*.

17. Robert L. Perlman, "Mouse Models of Human Disease: An Evolutionary Perspective," *Evolution, Medicine, and Public Health*, April 2016, eow014, doi:10.1093/emph/eow014.

18. Perlman, "Mouse Models of Human Disease."

19. Lydia Ramsey Pflanzer and Samantha Lee, "Our DNA Is 99.9% the Same as the Person Next to Us—and We're Surprisingly Similar to a Lot of Other Living Things," *Business Insider*, April 2018, https://www.businessinsider.com/comparing-genetic-similarity-between-humans-and-other-things-2016-5.

20. The ENCODE Project Consortium et al., "Perspectives on ENCODE," *Nature* 583, no. 7818 (July 2020): 693–698, doi:10.1038/s41586-020-2449-8.

21. Georgy S. Levit and Uwe Hossfeld, "Ernst Haeckel in the History of Biology," *Current Biology* 29, no. 24 (December 2019): R1276–R1284, doi:10.1016/j.cub.2019.10.064.

22. These two studies represent one of the first uses of gene editing in the treatment of human disease: Erica B. Esrick et al., "Post-Transcriptional Genetic Silencing of *BCL11A* to Treat Sickle Cell Disease," *New England Journal of Medicine* 384, no. 3 (January 2021): 205–215, doi:10.1056/NEJMoa2029392; Haydar Frangoul et al., "CRISPR-Cas9 Gene Editing for Sickle Cell Disease and β-Thalassemia," *New England Journal of Medicine* 384, no. 3 (January 2021): 252–260, doi:10.1056/NEJMoa2031054.

23. Barbara Rotstein and Achim Paululat, "On the Morphology of the Drosophila Heart," *Journal of Cardiovascular Development and Disease* 3, no. 2 (April 2016): 15, doi:10.3390/jcdd3020015.

24. R. Monahan-Earley, A. M. Dvorak, and W. C. Aird, "Evolutionary Origins of the Blood Vascular System and Endothelium," *Journal of Thrombosis and Haemostasis* 11 (June 2013): 46–66, doi:10.1111/jth.12253.

25. Thomas Browne and C. A. Patrides, *The Major Works*, Penguin English Library (Harmondsworth, UK; New York: Penguin Books, 1977), chap. 5.

26. Marja J. L. Kik and Mark A. Mitchell, "Reptile Cardiology: A Review of Anatomy and Physiology, Diagnostic Approaches, and Clinical Disease," *Seminars in Avian and Exotic Pet Medicine* 14, no. 1 (January 2005): 52–60, doi:10.1053/j.saep.2005.12.009.

27. Gail A. Van Norman, "Limitations of Animal Studies for Predicting Toxicity in Clinical Trials," *JACC: Basic to Translational Science* 4, no. 7 (November 2019): 845–854, doi:10.1016/j.jacbts.2019.10.008; Thomas Hartung, "Look Back in Anger—What Clinical Studies Tell Us about Preclinical Work," *ALTEX* 30, no. 3 (2013): 275–291, doi:10.14573/altex.2013.3.275.

28. J. Hodgson, "ADMET—Turning Chemicals into Drugs," *Nature Biotechnology* 19, no. 8 (August 2001): 722–726, doi:10.1038/90761.

29. Miller, *Drugged*.

30. Miller, *Drugged*.

31. Hartung, "Look Back in Anger"; Van Norman, "Limitations of Animal Studies for Predicting Toxicity in Clinical Trials."

32. Hinnerk Feldwisch-Drentrup, "New Clues to Why a French Drug Trial Went Horribly Wrong," *Science*, June 8, 2017, doi:10.1126/science.aan6949; Richard J. Miller, "Far FAAH Out," *The Keys to All Mythologies: Science, Medicine and Magic* (blog), May 2019, 363, http://richardjmiller.org/2019/05/04/far-faah-out/.

33. Channabasavaiah B. Gurumurthy and Kevin C. Kent Lloyd, "Generating Mouse Models for Biomedical Research: Technological Advances," *Disease Models & Mechanisms* 12, no. 1 (January 2019): dmm029462, doi:10.1242/dmm.029462.

34. Rebecca D. Hodge et al., "Conserved Cell Types with Divergent Features in Human versus Mouse Cortex," *Nature* 573, no. 7772 (September 2019): 61–68, doi:10.1038/s41586-019-1506-7.

35. Aysha Akhtar, "The Flaws and Human Harms of Animal Experimentation," *Cambridge Quarterly of Healthcare Ethics* 24, no. 4 (October 2015): 407–419, doi:10.1017/S0963180115000079.

36. There are some drugs that have been found to treat the symptoms of diseases like Parkinson's disease (PD) and Alzheimer's disease (AD), but I would argue that these weren't really developed through animal experimentation. L-DOPA, for example, is a great help for many PD patients but its use came from observations of human pathology and testing. For AD, the cupboard is pretty bare. Cholinesterase inhibitors provide very small improvements in cognition in some patients, but again, their development followed observations of the degeneration of forebrain cholinergic innervation in human pathological samples. Very few drugs that affect the nervous system were developed primarily through animal experimentation, although it has happened on occasion.

37. Gerald L. Posner, *Pharma: Greed, Lies, and the Poisoning of America* (New York: Avid Reader Press, 2020).

5

Fear and Trembling

Nature is the art through which God made the world and still governs it. The art of man imitates it in many ways, one of which is its ability to make an artificial animal. Life is just a motion of limbs caused by some principal part inside the body; so why can't we say that all automata (engines that move themselves by springs and wheels as a watch does) have an artificial life? For what is the heart but a spring? What are the nerves but so many strings? What are the joints but so many wheels enabling the whole body to move in the way its designer intended?

—*Leviathan*, Thomas Hobbes

Standing on a hill looking through a pair of binoculars, you view the countryside around you. On the top of another hill nearby is an elderly man. The man is standing over a pile of wood. Lying on the pile of wood is a younger man who is bound with ropes. The old man leans over him. His hand grasps a sharp knife raised above his head, showing his intention to strike and presumably kill the young man. It is clear to you that this is some kind of ritual murder. However, the elderly man seems to be conflicted about what he is doing. He doesn't want to strike. His face is contorted in agony and his body is trembling all over. He appears to be in the grip of intense feelings of fear and anxiety. It is not that Abraham has told you this directly. It is merely that observing his behavior allows you to construct a reasonable interpretation of the events unfolding before you. Abraham is about to kill his son Isaac. He is in turmoil, his emotions are difficult to control, and, as a result, he is trembling with passion. As you subsequently observe, Abraham relents and doesn't kill his son. Perhaps if you ever have the opportunity of meeting him in person, he may explain to you exactly what had happened and how he felt about it. Nevertheless, you have been able to intuit many of the key features

The Rise and Fall of Animal Experimentation. Richard J. Miller, Oxford University Press. © Richard J. Miller 2023.
DOI: 10.1093/oso/9780197665756.003.0005

of the narrative merely by observing Abraham's behavior. This is what usually happens.

When we want to understand what other humans think, we rely on two main sources of information: what they tell us and how they behave. But when we deal with animals, they can't tell us why they do things or how they feel. They don't share our language. And, anyway, as Wittgenstein (again!) said, "If a lion could speak, we wouldn't understand him."[1] I don't think Wittgenstein meant that we wouldn't understand the lion's language in the same way that an Englishman might not understand French or Spanish. Rather, the lion's entire frame of reference is not the same as that of a human; the lion views the world in a completely different way and would express himself using a completely different set of terms that we wouldn't necessarily comprehend. This raises the important question as to how we understand the mental world of animals. The way we think about them scientifically and ethically will surely depend on such an understanding. If an animal is a completely unfeeling machine, as Descartes and his followers would have had us believe, then we might generally act in one way, but if we think the animal has access to an entire repertoire of conscious experiences, then our actions might well be entirely different. Furthermore, given the way our cognitive processes regulate so many aspects of our physiology, understanding the mental processes of animals is essential if we want to evaluate how appropriate they are as experimental models for humans. Understanding animal minds, then, is a key piece of information we need to consider when deciding how to deal with animals from the point of view of science or any other type of interaction we might have with them.

But how can we understand the mental life of an animal? If we view an animal trembling, can we say the same thing as we might say of Abraham, that the animal is displaying a conscious response to feelings of fear and anxiety? Or is the response just a mechanical reflex devoid of any mental ideation and conscious reflection? It is common for humans to view animals from a completely anthropocentric standpoint and attribute all kinds of human emotions to them—and also, critically, the lack of them. But is this justified? Animals have emotions, no doubt, but is it reasonable to couch these in entirely human terms?

Of course, just as we interpret Abraham's feelings from his actions, we all surely do the same thing with animals as a matter of course. An example from literature: In Thomas Pynchon's novel *Against the Day*, Pugnax the dog is observed with his nose in a copy of Henry James's novel *Princess*

Casamassima on which the narrator comments, "He had learned with the readiness peculiar to dogs how with the utmost delicacy to turn pages using nose or paws, and anyone observing him thus engaged could not help noting the changing expression of his face, in particular the uncommonly articulate eyebrows, which contributed to an overall effect of interest, sympathy and— the conclusion could scarce be avoided—comprehension."

The problem of "other minds," that is, understanding the mental processes of another human, is a very hard philosophical and scientific nut to crack, let alone understanding the mind of an animal. Consciousness has been defined by philosophers as a "hard problem" but, as in Thomas Browne's words, "What songs the sirens sang or what name Achilles assumed when he hid himself amongst women, though puzzling questions, are not beyond all conjecture."[2] So, we must attempt to understand consciousness if we are going to know how to appropriately interact with animals. In particular, as far as science is concerned, how does our appreciation of animal minds influence the way we consider using them as translational models for investigating human disease, because virtually every aspect of a human disease is affected by our state of mind? This ranges from our appreciation of pain, to whether we are depressed and anxious, and even to things like our immune responses to pathogens. The power of the mind to affect physiology is clear from everyday occurrences like vasovagal syncope—the sight of blood and gore in some people triggers activation of the vagus nerve, a fall in blood pressure, and even, perhaps, loss of consciousness. This is old news. But today we know of many previously unsuspected effects of this type. Activation of the vagus nerve and secretion of its neurotransmitter acetylcholine can also act on receptors expressed by cells of the immune system such as macrophages and so regulate inflammatory responses. The field of neuroimmunology studies such interactions, bringing together the mind and the immune system, two areas of biology that used to be thought of as being completely independent, but in reality, they are anything but independent and constantly influence one another. One might also consider a phenomenon such as a conversion disorder, in which patients can present with extremely florid symptoms including seizures and paralysis but without the neuropathology that normally accompanies such states. Here the mind's eye, responding to some trigger such as a traumatic memory, initiates and drives the syndrome. Clearly, the mind can powerfully influence the manifestation of symptoms. If minds are uniquely human, how do we factor such considerations into our use of animals to accurately reflect human disease? Rats and mice are the animals that are most often used as experimental models of humans, so it is particularly important to know how and when this is appropriate. What is the mind of a mouse really like? If we are going to use animals like mice as models of humans, then just how much like humans are they?

We all grow up hearing stories about mice (or rats, guinea pigs, rabbits, squirrels, and other small furry animals) who are very human indeed. Someone like Beatrix Potter was expert at showing us mice who were very nearly human, attired in human clothes with human behaviors like drinking chamomile tea, going shopping, and being engaged in every conceivable type of human activity.[3] Really, Beatrix Potter's animals are tiny humans who have some animal attributes. They inhabit a netherworld that hovers between the human and the nonhuman. They are human enough for us to understand them and their motivation for doing things but animal enough for us to be intrigued by some of the things they do that could be suggestive of their animal "otherness" (Figure 5.1). As Susan Orlean has written about animals,

Figure 5.1 Hunca Munca cleans up.
Image from Alamy.

"They seem to have something in common with us, and yet they're alien, un-knowable, familiar but mysterious."[4]

Fictional animals are also sometimes our equals in many respects. Indeed, some of them are clearly more intelligent and moral than we are, including Jonathan Swift's Houyhnhnms or Pierre Boulle's apes, but here again, their intelligence and their moral compass are just another version of human intelligence and morality rather than something that might be uniquely their own. Plutarch, writing in the first century CE, described the life of Gryllos, one of the members of Odysseus's crew who was turned into a pig by the sorceress Circe and didn't think that becoming human again was something he really wanted to do because he had concluded that pigs were superior. Gryllos mentions that both male and female animals are brave and possess innate skills; he praises their abstinence from sexual perversion and their absence of gluttony and avarice. Overall, Gryllos attacked the Stoic argument denying reason to animals and convinced Odysseus of the moral superiority of many animals over humans.

In reality, however, we have very little evidence that mice really think in the way Beatrix Potter describes. Mice don't talk and don't have facial expressions and other behaviors that allow us to interpret the landscape of their inner emotions as we would another human being. Or do they? The study of animal behavior has advanced greatly over the last few years, and animals like mice are seemingly more and more complex every day. No, they don't dress up in clothes like ours or do many of the things humans do, but 85% genetic similarity does get you something in terms of shared core physiology and behaviors, as we have seen from the study of things like the cardiovascular system. Let's discuss some of these recent revelations. First of all, did you know that mice can smile? We are used to seeing different types of behaviors displayed by monkeys, dogs, and cats and inevitably intuit their thoughts and emotional states from our observations. But how about mice, the animals that are the most commonly used research model for humans? Considering how important human facial expressions are as a way of communicating with other individuals, the lack of this capacity in mice has certainly been one reason we have found it difficult to decipher their state of mind. But as things turn out, these views were based on the fact that we just didn't know how to measure the repertoire of mouse facial expressions properly, not that they didn't have them. A recent set of studies described in the journal *Science*[5] used a sophisticated machine learning algorithm for examining the faces of mice when they were put into seven different circumstances such as being given sugar syrup, which they like; an electric shock, which they don't; or other situations where they might experience fear, anxiety, and so on. The results were extremely clear. Each of the different circumstances was associated with different patterns of facial expression. These don't look like human facial expressions, but then they are mice after all. For example, like dogs and cats, mice do a lot with their ears and can curl them up when they feel "pleasure" associated with drinking sucrose. This is the kind of thing that helps a mouse express itself. At the same time that the mice were displaying different behaviors and associated facial expressions, the experimenters recorded the electrical activity of their brains and observed that nerves in an area of the brain called the insula, which is known to be associated with emotional responses in humans, fired different patterns at the same time. These revelations now mean that scientists can use mouse facial expressions in the future to better understand their responses to different stimuli or in different situations.

If mice have a much more complex set of facial expressions than we previously thought, what about language, the other main way that humans communicate? It is clear that animals don't use language in the same way that humans do. Language allows us to use words strung together in innumerable combinations according to certain grammatical rules to indicate ideas and concepts of an abstract nature. Animals can't do that, as far as we know. On the other hand, animals certainly do use sounds to communicate. But here again we don't traditionally think that the sounds mice make allow them to communicate a very extensive range of meanings. Nevertheless, such views also turn out not to be true. The problem for humans is that the sounds mice make are not generally audible because they occur at frequencies in the ultrasonic range, which humans can't detect. A large study published in the journal *Nature Neuroscience* recently highlighted these abilities.[6] The investigators used machine learning software called DeepSqueak to compare more than 111,000 individual vocalizations, including those in the ultrasonic range, accompanying more than 32,000 examples of various behaviors among mixed groups of male and female mice in a specialized recording chamber. Male mice, for example, tended to make distinct sounds depending on whether they were fighting, fleeing, chasing females, or engaged in some other activity. Moreover, if the particular sounds were made artificially, mice who heard them would respond just as if they were interacting with another mouse. Clearly, mice have an entire collection of ultrasonic signals at their disposal that they use to communicate in a sophisticated manner, even if it isn't exactly language in the human sense of the word.[7] Like Dr. Doolittle, we need to learn to communicate with animals on their own terms. It is becoming more and more obvious that they have plenty to say to us.

Rodents also do other things that might remind us of humans. For example, rats like to play hide and seek. This interesting discovery was reported a few years ago in the journal *Science*.[8] To study this possibility, the experimenters designed a large "playroom" equipped with boxes or nooks for the rats or a human to hide in. When the rat was the "seeker," games were initiated by placing a rat in one of the boxes, which could be opened electronically by a hidden human. When the box was opened, the rat would emerge and immediately start looking for the human. When the human was found, the rat would be rewarded by being tickled on its stomach (yes, rats love being tickled![9]). When the rat was the "hider," the human would leave the box open while the rat jumped out and ran to one of its designated hiding places where it could be found by the human. During some of the experiments the scientists made recordings from the rats' brains and observed that a set of nerve cells in an area called the prefrontal cortex became active during the game, and the pattern of activity was specified according to whether the rat was a hider or a seeker. To quote one of the paper's authors: "Many scientists think this is trivial, but these are very complex behaviors because the rats assume different roles, follow rules, and even strategize about where to hide." He also noted that the rats appeared to be playing the game for "fun." When a rat found a researcher, it would seemingly actually jump for joy, what the scientists called a *freudensprung*, saying, "This is something that a lot of mammals do when they are having fun, including rabbits, lambs, and people." In addition, the rats often scurried off to a new hiding place after being found, extending the game, and postponing the reward of being petted.

These recent results are not the only ones to have indicated the sophisticated ability of rodents to strategize or engage in forms of complex social behaviors that remind us of humans. In another report, rats were put in a situation where they had to choose between receiving a piece of chocolate (another thing rats love that is rather human) or saving a fellow rat from drowning. There was an overwhelming tendency for the rats to choose the latter course of action, indicating that rats can also display something along the lines of empathic behavior.[10] Another series of experiments demonstrated that mice can "feel each other's pain." If you put two mice together and inflict a pain stimulus on one of them and then test the other "bystander" mouse, you will find that its pain threshold to mechanical (being poked) or thermal (hot water) stimulation has dropped; in other words, the bystander mouse has become more sensitive to nociceptive (painful) inputs. This effect can even be produced by exposing a bystander mouse to the bedding of the mouse who had experienced pain, indicating that a chemical mediator like a pheromone may be involved in transmitting this effect.[11] Experiments such as these strongly suggest that animals such as rats and mice are capable of empathic behaviors, the ability to share the affective state of others.[12]

Nevertheless, rodents do not share all behaviors with humans. A recent study asked whether gerbils appreciated art in the same way that humans do. The investigators prepared tiny versions of several masterpieces including Vermeer's *The Girl with the Pearl Earring*, Munch's *The Scream*, Klimt's *The Kiss*, and De Vinci's *Mona Lisa*, but in each case substituting a gerbil's face for the human subject. These were then placed in a tiny art gallery complete with the appropriate furniture. When two gerbils (named Pandoro and Tiramisu) were admitted to the gallery, the investigators observed that they spent almost no time contemplating the paintings but a great deal of time chewing up the furniture and other objects (Figure 5.2).[13] This is not what humans would do. Beatrix Potter's mice wouldn't have behaved this way. In Schiller's words, "We become fully human beings when we contemplate beautiful works of art."[14]

Figure 5.2 Pandoro and Tiramisu chew up the furniture.
Image used with permission from Filippo Lorenzin and Marianna Benetti 2020.

This may seem like a trivial story that I am mentioning because I am trying to be funny, but I think it says something interesting about the differences between the minds of mice and humans. Aesthetics are a realm of mental activities that are unique to humans and are not shared, in any way that we understand, by other animal species such as rodents. Animals can certainly recognize pictures and music but do not show reinforcing preferences for these things like humans do. It is well known that animals such as pigeons can learn to distinguish different types of paintings—such as those of Picasso versus Monet—with great accuracy, but what cues they are using to do this and whether this has anything to do with their "appreciation" of the paintings is not clear.

In other words, mice aren't just little furry humans; they are mice. The tension between these two identities, knowing whether mice are really like humans in the Beatrix Potter fashion or whether they are something distinct, is at the heart of the debate about how useful an animal like a mouse may be as a reflection of a human in a scientific experiment. The answer, as we have discussed, is that it depends on the experiment.

Mice, then, are clearly capable of communicating with one another using facial expressions and sounds in a way that seems similar to humans in many respects. Obviously, the details of these communication systems differ when we compare one species to another, but the *core* of the idea is the same, just like the *core* idea of how the cardiovascular system works is the same in a mouse and a human. It is important for scientists to be able to say that a mouse's behavior is somehow comparable to that of a human because we tend to judge everything in the world from an anthropocentric perspective, and this becomes very important if we want to model different types of human mental activity in mice. However, we should also note that if we can say a mouse is really like a human in certain important respects, then this opens the door to considering another important possibility. Surely, we will have to grant the mouse the same kind of moral status that we would grant another person. If a mouse has sufficient features that are similar to humans, then perhaps a mouse should be considered to be a person in its own way and that "personhood" is not something that is unique to human beings. Nevertheless, just because a mouse can exhibit complex behaviors that are of the same type as a human doesn't mean that it is a Beatrix Potter mouse and what it experiences is truly human. This is clearly one of the problems we have when we interpret everything from an anthropocentric point of view. A mouse's behavior needs to be interpreted on its own terms.

And it's not only mammals such as rodents that may exhibit core behaviors that are somewhat analogous to humans. We can observe similar things if we look carefully up and down other branches of the evolutionary tree. Consider an octopus, for example.[15] The octopus family tree split off from that of humans some half a billion years ago. Being an invertebrate, having eight limbs, and living one's life in the watery depths may seem far away from the world of humans. But like a human, an octopus can behave in complex ways. For example, an octopus can learn to play a game with a ball and display empathic interactions with other octopuses. Note that in the world of invertebrates the octopus has a particularly large brain and has been appreciated for its high levels of intelligence, so perhaps it isn't surprising that behaviors such as these have emerged in this species. As it turns out, social interactions are highly regulated in octopuses, which usually display rather solitary behavior, except when it comes to mating. A recent paper described the effects of the drug methylenedioxymethamphetamine (MDMA) on octopuses' behavior.[16] This drug, which is commonly known as "ecstasy" (XTC), is well known for promoting empathic behavior and sociability in humans. The authors wondered what the drug would do in an intelligent but evolutionarily distant animal like an octopus. To find out, the authors performed the following experiment. They constructed an aquarium tank with three chambers. In one side chamber they placed a novel object (a soccer ball). In the other side chamber they placed an octopus in a cage. They then took another octopus and placed it in the central chamber to see what it would do. They observed that normally the exploring octopus would avoid the octopus in the cage and spend much more time playing with the soccer ball. Even when the exploring octopus did spend time with the octopus in the cage, it wasn't very friendly toward it, perhaps just extending one limp wrist in the direction of the caged mollusk. Then the experimenters put MDMA into the water for 10 minutes. Now the octopus's behavior changed dramatically. Instead of avoiding the octopus in the cage, the exploring octopus spent much more time in its company and much less time with the soccer ball. What is more, the character of the interaction changed. Now the exploring octopus examined the caged octopus with a full-on eight-arm hug. In other words, the MDMA had a pronounced prosocial effect on the octopus's behavior, just as it does with humans. It seems, then, that the ability to exhibit social behavior does exist in the octopus, but it is normally suppressed, except during mating. The results suggest that some aspects of social behavior are not only ancient but also controlled by the same neurotransmitter systems throughout evolutionary history. It is known that MDMA works by interacting with a protein called SERT (serotonin transporter) that is found in the brains of humans and octopuses. The SERT protein controls the levels of the neurotransmitter 5-hydroxytryptamine (5-HT, also called serotonin) that is found in the synapses that connect certain nerve cells in the brain. The role of this system in controlling "sociality" seems to have been conserved throughout evolution, having the same kind of function in both a human and an octopus, suggesting that this is an aspect of human social behavior where the core idea may be present in other more ancient animals as well. And it's not just the octopus that shows complex behaviors of this type. It has recently been shown that the cuttlefish (*Sepia officinalis*), a closely related species, can exhibit "self-control, the ability to overcome immediate gratification in favor of a better but delayed reward. This is a vital cognitive skill that underpins effective decision-making, goal-directed behaviors and future planning."[17]

When you call somebody a "birdbrain," it isn't supposed to be a compliment. When you look at the size of a bird's brain, it doesn't appear to be very large. Moreover, it doesn't seem to possess much in the way of a cerebral cortex, the part of the human brain that we think is responsible for phenomena like consciousness. Nevertheless, if we examine the behavior of birds, some of them are highly intelligent. They can learn how to use tools to solve problems, something that is particularly true of corvids—birds like crows and ravens. Recently, birds have been shown to exhibit "altruistic" behaviors involving helping other birds remove tracking devices attached to them by researchers.[18] Some scientists think that birds have intelligence that rivals that of the great apes.[19] The latest research has shown that the top of the brain, or pallium, in these birds is packed with neurons, and the electrical activity of that part of the brain closely resembles what is going on when scientists attempt to map the brain correlates of consciousness in humans. These results were considered to be so important that *Science* magazine recognized them as one of the 10 top "scientific breakthroughs" of 2020.[20] In other words, just because something doesn't look human doesn't mean it can't exhibit attributes that are similar to, or perhaps we might say "functionally equivalent to," those of humans. Overall, our appreciation of avian intelligence continues to increase at a remarkable rate.[21]

These examples suggest that, at the very least, many animals possess a rich and complex internal psychological environment. Some of the things that they do clearly remind us of human behaviors. If we saw another human doing these things, we would make important assumptions about what that person was thinking, so why would we not conclude the same thing about a mouse (or an octopus or a crow)? Not to do so would be to ignore what clearly seem to be parsimonious and scientifically appropriate conclusions. Or should these behaviors be interpreted differently? Scientists traditionally interpret the behavior of animals according to opinions that arose at the end of the nineteenth century, such as those of Lloyd Morgan. "Morgan's canon" stated, "In no case is an animal activity to be interpreted as the outcome of the exercise of a higher psychical faculty, if it can be fairly interpreted as the outcome of the exercise of one which stands lower on the psychological scale." Behaviorism jettisoned any consideration of introspection or the possibility of animal minds, leaving only a consideration of their actions described in language that was purely "scientific" and devoid of empathetic references. Although the study of animals has moved on from such views, they remain influential. When a scientist encounters a mouse, it is usually in a small white cage where it has very few opportunities for displaying interesting behavior. Under these circumstances it is probable that your opinion of what this animal is capable of doing might be extremely limited. Moreover, even animal behaviorists have done a very poor job of decoding the complex behavioral possibilities of their animal subjects until very recently. The examples I have discussed above have all been published over the last couple of years. Yes, we have known for a long time that mice can exhibit different behaviors. Mice can be made to follow an operant schedule in a Skinner box. They can learn to press a lever to get another hit of heroin. They will learn to freeze in place or avoid areas of their home cage where they have been given electric shocks. Whatever the usefulness of these techniques, they have told us little about the behavior of a mouse *qua* mouse, and if we are going to make decisions as to the state of mind of such an animal, we must first have a clear picture of what they are capable of doing on their own terms, not merely as reflections of human experience. Again, as Wittgenstein said, "If one sees the behavior of a living thing, one sees its soul."[22] Here I don't think Wittgenstein was referring to some metaphysical soul of the type that Descartes might have considered, but rather the true essence of the personhood of the creature. We need to try and understand animals not as Beatrix Potter would have us understand them but as they understand each other and themselves. Even if animals display behaviors that look human, what do these things mean to the animal itself? This is key information if we are going to consider the appropriateness of an animal as a human experimental model. To answer such a question, we need to try and understand not only their consciousness but also their self-consciousness, what is referred to as meta-consciousness or consciousness about consciousness. Are the human-like behaviors that are displayed by animals really indicative of the fact that they are conscious entities, and if so, what kind of conscious entities? Moreover, even if that is the case, how much like a human mind is the mind of an animal?

Inside Animal Minds

The question of consciousness is a slippery one for most biological scientists to deal with. What exactly is it, and how do you attack it experimentally? If you ask the vast majority of neuroscientists about consciousness, they will reply that it is "something to do with the electrical activity of the brain." When lots of neurons in the brain fire electrical signals in unison then *voila!*—we have consciousness. But then, having said this, they will falter on exactly how the electrical activity produces consciousness and how different types of electrical activity produce different types of consciousness in different people, in animals, and so on. When treated in an unsophisticated manner, this kind of approach leads to the common science fiction trope that you could "upload" your consciousness to a machine or the internet—after all, it's just a lot of neurons firing electrical signals, right? A few years ago, I attended a performance of *Death and Powers: A Robotic Opera*, composed by faculty at MIT, in which the protagonist does indeed upload his consciousness to the internet to free himself from the "land of meat" and live forever.[23] In this kind of scenario, consciousness is just a lot of electrical activity, and, like a piece of music or a text, it can be recorded and transferred from one type of medium to another. However, in reality, things just aren't that simple. As we all know, what we think is very dependent on the bodies we inhabit and how they interact with the world. If our consciousness didn't have a body or was placed in the body of another human, an animal or a machine, presumably it wouldn't be the same consciousness any longer. Even if we could upload the experiences of other human beings and explore them ourselves, we might still have problems with doing the same with the mind of an animal, as implied by Wittgenstein's lion aphorism. Mind and body constantly interact and are codependent. One cannot change one and not the other. Indeed, an ever-increasing corpus of data has demonstrated how the activity of our internal organs, transmitted to our brains by nerves like the vagus, constantly helps to shape the specific consciousness of an individual—a process known as "interoception."[24] If the consciousness of a human was "uploaded" into an animal, it would no longer be the same consciousness.

The problem of how we view the internal mental landscape of an animal in comparison to how we describe ourselves was discussed in a famous 1974 essay by the philosopher Thomas Nagel entitled "What Is It Like to Be a Bat?"[25] Since that time the same kind of title has been used by many authors to frame a discussion on the nature of animal consciousness, resulting in a large number of essays entitled "What Is It Like to Be a [*fill in the name of your favorite animate or inanimate object*]?" At the moment, for example, I am reading an article called "What Is It Like to Be a Bee?"[26] It is not at all clear, based on our own experience, what the subjective world of another animal would be like or even if they have one at all. It presumably must depend on the physical nature of the animal in question—a human, a bat, a mouse, a bacterium. Some schools of thought attribute a degree of consciousness to virtually any material object (panpsychism), which raises some interesting issues, as pointed out by the front cover of the British satirical magazine *Private Eye* in 1966 following the first pictures of rocks on the lunar surface sent back to Earth by *Lunar 9*, the first unmanned spacecraft to land on the moon (Figure 5.3).

Nagel's essay and other philosophical writings have highlighted the idea that our conscious experiences revolve around the phenomenon of *qualia*, which express the essential subjective essence of a thing. Consider the color red. How would you describe redness to a person who had been blind from birth? Your concept of redness has qualities that result from deeply personal aspects of your conscious experience. This results from "what it is like to be you." Another person will have a qualitatively different experience depending on "what it is like to be them," although, considering they are human, we assume their experience would share some features that are similar to our own. But a bat or a bee (or a mouse or an octopus) is a completely different proposition, and yet, unless they are zombies or machines, they must have a world of experiences as is clear from the examples we discussed above. Nevertheless, the way that these animals understand the world is based on very different physical perspectives, so "what it is like to be them" is certainly not what it is like to be a human. Using color perception as an example, humans use three types of retinal cells called cone cells that detect red, blue, and green. However, mice, cats, and dogs only have blue and green cones. This means they have a much less extensive perception of color, perhaps something akin

Figure 5.3 Panpsychism? Front cover of *Private Eye* magazine following the first moon landing, June 1966.
Image used with permission from *Private Eye*.

to color blindness in humans. So, when looking at a color with red in it, a mouse will see something, but it won't be red—that's for sure!

Can we understand, even in principle, what that something is? Generally speaking, scientists fail to come to grips with the problem of the essential otherness of animals. Rather, and this I think is the real problem, a scientist who wants to use an animal in an experiment will think of the animal as if it was an incomplete version of a human rather than something else that has conscious validity on its own terms. There is a large literature dealing with the potential for animals to use human words and language. We are fascinated to see a gorilla or a crow that can recognize human words and utilize them effectively. It seems to me that these studies are misguided and teach us nothing. It is like an experiment that sets out to train a human being to fly like a bird. Initially he jumps out of a tree and flaps his arms to no avail, rapidly falling to the ground. Perhaps after years of training, his arm muscles and coordination have improved to such an extent that he can actually travel a few yards through the air. What does this prove exactly? That a human is an incomplete version of a bird? Animals, as we have discussed, have their own modes of communication and behavior that are precisely appropriate for their own species, and the fact that they do not use human language and cannot easily appropriate it is because it is entirely unnatural and un-necessary for their existence. As the American naturalist and writer Henry Beston said in 1928, "Animals are not bretherin, they are not underlings: they are other nations, caught with ourselves in the net of life and time, fellow prisoners of the splendor and travail of the Earth."[27] Animals, then, need to be understood on their own terms.

This is of great importance if we are going to use mice as models of humans in the case of any disease in which conscious experience is important, which is probably most of them. Pain, for example, is undoubtedly an important consideration in medicine, but can we use a mouse to model human pain? Do we know how pain is perceived personally by a mouse? And is that experience useful to us as humans if we want to use mice as experimental models? As we will discuss, scientists often use animals like mice in attempts to understand phenomena like pain. Scientists make measurements of animal "pain." But what is it we are measuring? If we aren't measuring the correct thing, then these experiments will not be useful, at least as far as translating results to humans is concerned, and our attempts to develop new drugs for pain will not be successful. Interestingly, there are drugs that treat pain in humans, but they were all obtained from Nature or by accidental off-label use of established drugs in humans. No novel drugs for pain have ever been developed using animal models. Perhaps there is a good reason for this. Perhaps human intuitions as to what constitutes pain in different animals is simply way of the mark. Indeed, as far as Nagel was concerned, qualia are beyond the ability of material theories of the brain to explain.

The biologist Jakob von Uexküll coined the term "*umwelt*," the idea that every organism responds to the surrounding world in different ways because it has a different sensory apparatus.[28] As von Uexküll expresses it, the worlds of humans and animals merge together in numerous ways to produce a "symphony of meaning and understanding." It is true that both humans and dogs have two eyes and a nose, and the basic mechanisms as to how these work are the same in both species. Nevertheless, dogs have many more odorant receptors than humans and so detect odors in the precisely same environment differently. As we have seen, there are also differences in the rods and cones in the eyes of the two species, so visual information is processed differently. The same applies to all sensory inputs as well as to the fabric of the brain that processes this information. Evidence to date suggests that dogs do not perceive visual illusions in the same way that humans do.[29] You and your dog may be in the same physical environment, but you will have your own specific interpretation of the world that differs from that of your dog. This seems reasonable. Every animal has evolved to occupy a unique biological niche. Your dog or any other animal, then, needs to be understood "on its own terms." The *umwelt* of a mouse and a human cannot possibly be the same for compelling biological reasons. There may be some overall similarities; however, when you want to use a mouse to develop a therapy that can be used on a human, then this is unlikely to be enough in many instances. The conscious experience of a mouse and a human cannot be the same, and this may be critical for attempts to use mice as human models.

However, be that as it may, most scientists do not care what philosophers like Nagel think. If there is a phenomenon to be measured, then surely science can measure it, explain it, and ultimately dominate and manipulate it. And so it is with consciousness. Consciousness studies are considered to be one of the more interesting (and challenging) areas of neuroscientific investigation these days, even by philosophers, who describe self-consciousness as a "hard" problem,[30] presumably to distinguish it from other "easy" problems that they have already solved, like "what is the meaning of life?"

These investigations are basically reductionist in nature. Study how the parts of the brain work in relation to conscious phenomena and as a result you will come to understand consciousness.[31] Some of this work is extremely sophisticated, and the results have been very interesting. Nevertheless, such investigations are always likely to suffer from the criticism known as Leibniz's Mill argument, articulated by the great eighteenth-century German philosopher Gottfried Leibniz as follows: "it must be confessed, moreover, that perception, and that which depends on it, are inexplicable by mechanical causes, that is, by figures and motions. And, supposing that there was a mechanism so constructed as to think, feel and have perception, we might enter it as into a mill. And this granted, we should only find on visiting it, pieces which push one against another, but never anything by which to explain a perception. This must be sought, therefore, in the simple substance, and not in the composite or in the machine." So, if we enter the brain and look around at all its different parts and see how they work together, will that be enough to explain self-consciousness or to answer questions such as "what is it like to be something"? It isn't easy to see how. Nevertheless, somehow the particular consciousness of each human or animal must result from the activity of its brain. Let's take a look at some of the types of experimental results that may give us insights into the way the activity of the brain and consciousness are related to one another.

Mechanisms of Consciousness

One thing that is certainly true is that by interfering with the substrate of the brain, we can greatly change an individual's conscious experiences. The "mother" of all these studies harkens back to experiments carried out by Roger Sperry and his many famous colleagues and students beginning in the 1950s and '60s.[32] At that time surgical interventions were sometimes employed to try and control severely incapacitating epilepsy (nowadays drugs are more often used to do this). The operations could be very successful in some instances and patients were much improved. Portions of the brain were sometimes removed or pathways connecting different parts of the brain severed in an attempt to reduce intractable seizures. In another population of patients an area of the brain had been destroyed due to a stroke. Once these groups of patients had recovered, they offered scientists a wonderful opportunity for studying the human brain. If a particular part of the brain was removed, destroyed, or severed, scientists could observe the consequences of this for different types of behaviors.

You might ask, for example, what happened to consciousness and related cognitive phenomena in these injured individuals? The scientists involved in the original studies concentrated on patients who had had a nerve pathway called the corpus callosum severed as a treatment for epilepsy. The corpus callosum is a huge bundle of nerve fibers that runs just beneath the cerebral cortices and connects one side of the brain to the other, allowing the two sides of the brain to communicate with one another. So, how did people behave if the corpus callosum had been severed? At a first pass, they appeared to be quite normal. They lived perfectly normal lives; their IQ or job performance wasn't altered in any obvious way. Ice cream was good, and pain was bad. However, if the patients were subjected to precise neuropsychological testing, the results were very strange. In humans, visual information that enters one eye travels through nerve pathways to the cortex of the brain on the opposite side of the body where it is processed, ultimately resulting in a mental image. If a picture of a cat is presented to either your left or right side and you are asked what you saw, you will say "cat." This is interesting because the ability to use language resides predominantly in your left cerebral hemisphere, which would receive images from the right-side presentation but not from the left. So, somehow this information must normally be shared between the two sides of the brain. But this isn't what happens if you present these pictures to a patient with a severed corpus callosum. If presented to the right eye the patient would say "cat," but if presented to the left eye the patient would say that they didn't see anything. On the other hand, if the patient was asked to answer nonverbally, say, by pointing to something like a picture, the cat would be recognized on whichever side the image was presented to. In other words, the patient now behaved as if they had two separate brains that didn't communicate with each other properly. In another experiment a card with an instruction on it was presented to the left side and so to the right "nonverbal" cortex. The patient would follow the instruction to stand up or make some other movement. However, if the patient was then asked why they had done it, the verbal left hemisphere didn't remain silent. It would usually say something like "I needed to go to the bathroom." In other words, it would provide a feasible narrative, making things up or guessing so as to try and make sense of reality. These experiments clearly showed that when the brain was divided in this way, in many respects it behaved as two individual entities, two minds, or two consciousnesses. Similar kinds of experiments have been carried out in animals like cats and monkeys in which the two sides of the brain have been experimentally separated and the results have led to similar conclusions when outcomes like the learning of specific tasks are measured.[33]

There are plenty of other weird results that have come from the study of the relationship between the physical substance of the brain and conscious experience. For example, when visual information reaches the brain, it is processed first by the primary visual cortex or V1 area. Suppose this area is severely damaged due to a stroke. One result is that the patient is functionally blind, and if asked whether they can see an object that is presented to that area of the brain, they will reply that they cannot. For example, let us say they are presented with a picture of horizontal or vertical stripes; they will say they can't see any stripes. Now ask them to just go ahead and guess. Amazingly, they are correct around 90% of the time. In other words, somehow they were "seeing" the stripes even though they weren't conscious of them. This phenomenon is called blindsight and is certainly very hard to explain. Whatever the explanation, it is clear from the split-brain, blindsight, and many other studies that if you interfere with the material substance of the brain, aspects of how we perceive reality can change in extraordinary ways. Perhaps this is at least getting our foot in the door if we are trying to approach the phenomenon of consciousness from a traditional experimental scientific point of view, and amazing results like these continue to be obtained by studying very unusual human patients.[34]

An Acid Test for Consciousness

However, to my mind, there are other kinds of experiments that have been carried out by far larger numbers of people that strikingly illustrate the relationship between the activity of the brain and conscious experience, so I would like to spend a few moments discussing them. Not only have these experiments been carried out by tens of millions of people, but this has been going on for thousands of years. I first became aware of hallucinogenic drugs in the 1960s when I was a teenager. Clearly, I was not alone. Taking drugs like LSD was one of the cardinal features of student culture in the 1960s, and that included a lot of people. Reports of experimentation with hallucinogenic drugs are often truly remarkable for many reasons but particularly from the point of view of consciousness and its mechanisms. Under the influence of a drug like LSD the relationship between one's conscious experience and objective reality is subject to changes that can be quite remarkable. The entire experience of who one is and what one's consciousness consists of may change radically. Importantly, considering the role of qualia in our subjective view of the world, the appreciation of things and their meaning for us is altered. What is red? The quale of red, of redness itself, is not what it was. And, of course, much of this is ineffable and hard to explain through ordinary discourse. The arts rather than the sciences are probably the best way to describe these experiences, and many have tried to do so, ranging from writers like Aldous Huxley, to musicians like Jefferson Airplane, to artists like Victor Mancuso.

But consider what is going on here. All one is doing is ingesting a vanishingly small amount of a pure chemical substance. Around 100 μg of LSD is enough to produce radical changes in conscious experience for most individuals. Taking hallucinogenic or "psychedelic" drugs produces an experience that is so unusual that many people who experience it are forced to reach into the realm of transcendence and metaphysics when searching for adequate descriptors. In ancient times, naturally occurring versions of these substances formed the basis of many religious practices throughout the world, and may even have been what prompted some cultures to develop religion in the first place, according to entheogenic theories.[35] In the 1960s Timothy Leary, the high priest of hallucinogenic drug culture, inaugurated what was in fact a new religion, widely adhered to by many young people, that fused the psychedelic drug experience with aspects of Tibetan Buddhism.[36]

If we are interested in the phenomenon of consciousness, then the effects of psychedelic drugs should also interest us greatly. They might be very useful tools for exploring the nature of consciousness in humans and animals. Many scientists have already thought about this possibility. For example, in 1980, Stanislav Grof remarked, "It does not seem to be an exaggeration to say that psychedelics, used responsibly and with proper caution, would be for psychiatry what the microscope is for biology and medicine, or the telescope is for astronomy. These tools make it possible to study important processes that under normal circumstances are not available for direct observation."[37]

From the hard science point of view, we know quite a bit about psychedelic drugs. Drugs like LSD have a very specific site of action in the brain.[38] It is a receptor protein called the 5HT2A receptor. 5HT, also known as serotonin, is an important neurotransmitter (chemical messenger) in the brain that we discussed above when we considered the effects of XTC on octopuses. Nowadays, we have all probably heard of drugs used for treating depression called specific serotonin reuptake inhibitors (SSRIs). Now we see that 5HT is also involved in the effects of psychedelic drugs. The normal function of 5HT receptors in the brain is to enable 5HT to carry out its job of communicating information from one nerve cell to another. There are fourteen different types of 5HT receptors, each specialized for carrying out different aspects of 5HT-mediated neurotransmission. Interestingly, most 5HT2A receptors are localized in nerve cells in the cerebral cortex, which, we think, has an important role to play in enabling whatever neural magic is responsible for the emergence of consciousness. Looked at in another way, a drug like LSD is a wonderfully potent and precise tool with which scientists can probe the underlying fabric of the conscious brain as noted by many researchers.

Indeed, it seems to me that hallucinogenic drugs potentially provide us with a way of really investigating qualia, one of the aspects of conscious experience that is most intractable to experimental investigation. This is because one of the key aspects of the psychedelic experience is a change in qualia. The personality of redness is now something different; the feeling of the sun on your skin is now a different feeling. Moreover, hallucinogens also often produce the phenomenon of synesthesia, in which conscious impressions become transferred from one sensory modality to another.[39] There are many different examples of this phenomenon. Sounds elicit colorful qualia, colors elicit sounds, sounds elicit tastes, and so on. In one of the most common forms, known as grapheme-color synesthesia, letters, numbers, and words evoke the experience of color. The phenomenon of synesthesia does occur naturally in a fair number of people, but it is common as one aspect of the psychedelic experience. A synesthete can sometimes even produce a meaningful experience of a color or other sensation merely by thinking of an appropriate trigger rather than experiencing it directly. Both synesthesia and hallucinogenic drugs modify the self-conscious phenomenon of qualia. Hence, investigating the mechanism of hallucinogenic drug action may be informative in our understanding of consciousness.

Given their extraordinary potential, it might appear strange that psychedelic drugs have been used very little over the years for the scientific exploration of consciousness. But there is an explanation for this—they are illegal. Richard Nixon placed LSD and other psychedelic drugs on Schedule 1 of the Controlled Substances Act in 1970, meaning that the government viewed them as incredibly dangerous and without any medical utility whatsoever. At that point all official research activity on hallucinogenic drugs completely fell off the map. In the last few years, however, things have started to change and research on these substances has started to pick up again using newer experimental methods that were unavailable in the 1960s.

With that in mind, how might contemporary scientists investigate the effects of drugs like LSD? Clearly we would like to have a real-time picture of what happens within the human or animal brain under circumstances in which states of consciousness are altered. Perhaps that would give us a way of starting to approach the "hard problem" of self-consciousness. Newer technologies that were not yet available in the 1960s now allow us to obtain pictures like this. The nerves in the brain work by transmitting electrical signals and, as we have discussed, communicate with one another using the chemical signals called neurotransmitters. 5HT, for example, is the neurotransmitter system that is responsible for producing the effects of hallucinogenic drugs. One can in principle measure the electrical activity of nerves in the brain directly by inserting recording electrodes into the appropriate place. However, it is deemed unethical to do this kind of thing to humans unless they happen to be undergoing a surgical or other intervention anyway, when, if you have obtained the correct permission, you may be able to sneak an electrode or two into the brain while the procedure is in progress. But this doesn't happen very often, and the conditions are somewhat inflexible. It would be much more useful, generally speaking, if one could do this "noninvasively," so that the activity of the brain could be assessed from the outside without having to dig into it. This is possible nowadays by using what are called "live brain imaging" techniques, particularly a method known as functional magnetic resonance imaging (fMRI), together with a method known as blood oxygen level dependent (BOLD). During the normal activity of the brain, when nerve cells release neurotransmitters, they increase local blood flow. Blood contains the molecule hemoglobin, which has an iron atom at its center. This atom can be associated with oxygen. The BOLD method measures local blood flow by assessing the oxygenated state of hemoglobin. So, it is not neuronal activity per se that is being directly measured but an effect that can be used to assess local nerve activity. The use of these techniques has completely revolutionized the study of human and animal brains when they are going about their business. Moreover, this kind of brain imaging is also a very powerful method of studying what happens to the activity of the brain in the context of pathology such as Alzheimer's disease, stroke, or other disorders. However, from the point of view of our present discussion, some of the most interesting results to have been obtained concern what happens when the brain isn't doing much of anything. Of course, if you are alive, your brain is never actually doing "nothing." In this instance what one means is that it isn't "on task," that is, paying close attention to a particular problem. Imagine that it is one of those times when you are just involved in self-reflection and exploration of your personal internal world. The interesting observation that has been made is that under these circumstances, a group of structures in the brain shows strong activity that is correlated—that is, "connected" in a network. This doesn't necessarily mean anatomically connected, as would be described by classical neuroanatomy, but connected in terms of correlative patterns of brain activity as measured by BOLD (or by other similar methods for functional imaging). This set of connected structures was named the default mode network (DMN).[40] It is a group of brain regions characterized by high volumes of blood flow and energy use.

Once it had been discovered, the concept of the DMN was immediately seized upon as a useful construct for discussing many types of normal and pathological forms of brain activity. One area of interest is the idea that it represents something akin to a neurological correlate of the "ego," or at least some sort of representation of subjective self-reflective mental activity. Such an idea has been suggested as being a useful starting point for framing questions about things like self-consciousness. It has been widely reported that many people who take hallucinogenic drugs experience a phenomenon known as ego dissolution or ego death, a loss of their sense of self and a growing feeling of transcendence and connection with Nature—broadly defined. Of interest, then, is what happens to the activity of the brain when an individual takes a hallucinogenic drug? Brain imaging studies have shown that the feeling of ego dissolution is correlated with a reduction in the connectivity of the DMN and increased connectivity between other areas of the brain.[41] The authors of these studies describe the brain as becoming more open or "entropic."[42]

The scientists concerned with this work have offered an interesting hypothesis that the brain operates at the level of two types of consciousness. The type of consciousness associated with the stabilization of the DMN is called secondary consciousness, or phenomenal consciousness. This is what we have referred to above as self or meta-consciousness and may represent an evolutionarily more advanced brain state associated with the potential for self-reflection. Primary consciousness, on the other hand, the type that is enabled by hallucinogenic drugs, meditation, or mystical religious thinking, is perhaps a more ancient brain state and is associated with different, more extensive patterns of correlated brain activities.[43] This theory, then, allows for the existence of different kinds of consciousness, not all of which are metacognitive, and, perhaps, opens the door to an understanding of some of the conscious processes that occur in some animals rather than humans.[44] It should be pointed out that in some ways brain imaging measurements do not get us any closer to understanding aspects of consciousness that would allow us to answer Thomas Nagel's question.[45]

We have seen that just like human behaviors, mice have sets of behaviors through which we might judge them. They produce facial expressions and sounds, play games, exhibit empathy, and, the more we look into it, normally display many behaviors that, if they were human, we would attribute to certain conscious states of mind. What happens if we view animals from the point of view of experimental neuroscience as provided by brain imaging studies? Animals such as great apes certainly display phenomena such as split-brain behaviors and blindsight. Apes have the equivalent of a DMN whose destabilization is associated with attention to tasks in a similar way to what is observed in humans.[46] Even mice have what appears to be the rudimentary equivalent of a DMN.[47] Hence, animals do have structural components that may be linked in ways that in humans seem to correlate with different aspects of consciousness, including self-reflection.

As it turns out, drugs like LSD have been given to an enormous number of animals but mostly in the 1950s and '60s in experiments that were not very well designed by our current standards so that it is difficult to assess the results from a contemporary experimental perspective.[48] Spiders were reported to spin rather free-form (one might even use the word "groovy") looking webs, fruit flies provided LSD on blotting paper (extremely authentic from that point of view) had disturbed vision, dolphins swam upside down, mice walked backwards (sometimes), monkeys had visual problems, and a poor elephant called Tusko in the Chicago Zoo had a seizure and died. Overall, it was difficult to find a clear behavioral response in an animal like a mouse that we could use to judge a mouse's response to hallucinogenic drugs. Eventually one was found. The major behavioral phenomenon used for examining the effects of hallucinogens in mice these days is the "head twitch," which seems like a pretty disappointing substitute for all the acres of transcendence attributed to the effects of LSD in humans.[49] Beyond overt behavior, however, when scientists have applied the latest neurobiological techniques to examining the effects of LSD in animals like mice, it does appear that a lot is going on, and given the fact that there are 5HT2A receptors in the brains of virtually all animals, this isn't surprising. Most recently the effects of hallucinogenic drugs have been examined on the properties of neurons in the V1 area of the visual cortex of mice and showed extremely large changes in the behavior of nerve cells in this area,[50] something that presumably must have altered the way mice process visual-sensory information. So, a great deal seems to be happening in brains of mice when they receive hallucinogenic drugs, but how this affects the quality of their sensory experience has been hard for us to judge. Nevertheless, given the profound effects of psychedelic drugs on so many aspects of human consciousness, their effects in animals may act as indicators as to the connections that exist between humans and animals. It should also be pointed out that many of the interesting, sophisticated behaviors displayed by mice, such as the wealth of facial expressions and sounds they use while communicating, which we discussed above, have only been discovered very recently, and the effects of hallucinogens or many other interesting experimental interventions have not been investigated on them as of yet. It may well be that hallucinogenic drugs will affect these behaviors in interesting and significant ways.

A Time for Reflection

Despite all the fascinating and technically sophisticated current research into the subject of consciousness, it remains a phenomenon where it is difficult to answer critiques such as those that arise from the thinking of Nagel and Leibniz. How the workings of the nuts and bolts of the brain machine will ever allow us to answer such questions in terms of the expectations of traditional scientific investigations remains unclear. Indeed, it isn't even clear yet, as we move further and further away from humans, what aspects of animal consciousness are truly analogous to what goes on in humans. Animals may have structures that look like a DMN, but are there other kinds of evidence that animals really have a form of self-reflective consciousness? For example, can animals really recognize themselves? Attempts to attack this issue have involved a procedure known as the "mirror test," devised by a scientist named Gordon Gallup back in 1970.[51] It's a very low-tech affair by today's standards but an interesting idea, nevertheless. Put a mirror in front of an animal and see what happens; does it recognize the reflection as itself? Depending on the animal, it will usually react by demonstrating some kind of interest. This may well be initial hostility: who is this other animal and what is it doing in my territory? After some time, however, when it becomes clear that no amount of screaming and jumping up and down is going to make any difference and the new animal is not going to go away, the animal may well get used to the situation and ignore it or even, interestingly, see what it does when provoked in some way. All these things aren't that surprising. The clever part of the experiment is what comes next. Unbeknownst to the animal, the experimenter will paint a spot on it, say, on an ape's forehead, where it can't be seen by the animal itself. Now what happens when the animal looks in the mirror? The reflected animal in the mirror now has a red spot on its forehead. What would I do under these circumstances? If I think that the reflection is me, I may well poke at my forehead and examine the spot. The idea, then, is that this will only happen if the animal can recognize its reflection in the mirror as itself. If it does, it is deemed to have passed the mirror test and possess something akin to self-consciousness. When this test was given to different nonhuman species, it was clear that chimpanzees and bonobos, our nearest genetic relatives, passed quite easily. They got the hang of it with alacrity. On the other hand, gorillas and different kinds of monkeys fared less well. To test other kinds of animals, the technical details for giving the test needed to be changed, but, as you might expect, dolphins passed it easily. Killer whales apparently did well (hard to imagine doing that experiment), and depending on who you ask, elephants also passed the test: indeed, a much-discussed elephant named Happy in the Bronx Zoo performed so well on the test that legal scholars have argued that she should be considered as a "person" under the law.[52] As far as birds are concerned, corvids, that is, birds like crows, who are famous brainiacs, were reported to have passed. Humans passed as well, of course, but only once they reach the age of two or so. Lots of other animals like dogs, cats, rats, and mice flunked the test. While all of this may be roughly in line with our general view of animal "intelligence," questions have recently arisen about the entire enterprise because of experiments demonstrating that a small fish called the cleaner wrasse (*Labroides dimidiatus*) reportedly aced the test.[53] Now, scientists aren't sure what mirror test results mean. Some investigators think that either self-awareness is much more widely found in the animal kingdom than previously imagined or the test isn't really measuring self-awareness. The experiment does seem like a good idea, but the results are odd. If an animal clearly passes the test, then I think this is evidence for some kind of self-consciousness. But given what we know about animals, why wouldn't a gorilla, dog, cat, or mouse pass if a small fish can pass? To be honest, with the way these experiments are performed and the way the data are obtained in many species, it isn't totally clear what is being measured and whether the animals are passing the test or not. Somebody even tried it on ants, who did okay—which seems rather strange.[54] Anyway, it is important to know about the mirror test because it has been discussed widely by the scientific community when considering theories of consciousness in animals, and the results are certainly interesting and represent another piece of evidence suggesting that something akin to self-consciousness is present in many nonhuman species.

Translating Consciousness

Now it is time to return to the original question we asked at the beginning of this chapter: how do all of these considerations help us to decide whether animals are useful for answering questions about human diseases? As we discussed in the previous chapter, animals are routinely used in biomedical research as stand-ins for every kind of human disease in the hope of obtaining results that can be usefully applied to humans. In some instances, starting with Ehrlich's use of rodents and rabbits in the development of Salvarsan, animal models have certainly been shown to be helpful. This is clearly true for at least some infectious diseases. And there are certainly other situations in which animals might model a human disease effectively enough to be helpful in the search for novel therapies. But there are also an enormous number of cases where this isn't really possible, and considering the potential influences of mental activity on physiology and even on things like our immune responses, why should it be? As discussed in the previous chapter, modeling diseases in mice becomes particularly difficult in cases where conscious experience is of central importance in the disease and represents aspects of its primary symptomology. This is true, for example, in the case of some neurologic and many psychiatric diseases. Diseases like depression and schizophrenia are primarily disorders of human conscious experience. One wonders, therefore, how one would model such a thing in a mouse, which, as we have discussed, presumably has conscious experiences that are uniquely mouse-like and not just watered-down versions of human experiences? This has taxed the minds of scientists greatly, resulting in hundreds of thousands of scientific publications on the subject at a cost of billions of dollars, so it is interesting to examine these particular studies further to see what they have come up with.

Let us consider what is known about unipolar major clinical depression, an extremely debilitating disorder that affects millions of people throughout the world including members of both sexes (although women are more susceptible to it), as well as different races and cultures. A scientist trying to approach this problem would argue like this: What are the major symptoms of depression in humans? Can I find a way of producing these symptoms in mice? Then, can I find a drug that abolishes these symptoms in mice? Now, give this drug to humans who have these same symptoms and presumably they will be cured. This is what Ehrlich did when he developed Salvarsan, but in that case the endpoint he wanted to measure was infection with *Treponema pallidum*, something that can be clearly demonstrated in animals. In most respects *T. pallidum* infection of rodents appears similar to infection in humans. Hence, mice can be used effectively as models for at least some bacterial infections in humans.

What are the symptoms of depression in humans? Depressed mood for an extended period of time would obviously be one of them. Another key behavioral feature is what is known as anhedonia—an inability to enjoy things that you previously found pleasurable. Low self-esteem, feelings of worthlessness, and an inability to concentrate are all frequently symptoms of depression, as are sleep disturbances (either too much or too little), feeling weak or listless, and lacking energy. Inappropriate weight gain is also often associated with depression in humans. In really severe circumstances, someone may even have suicidal ideation. How are mice used to model these things? This is clearly something that is a lot more involved than *T. pallidum* infection. So, what have scientists come up with?[55] For a long time now one popular test has involved putting mice into a bath full of water. Mice don't like to swim and after some time spent splashing around, they will eventually give up. If you put them into the cylinder again a day later, you will observe that they spend a lot of time just floating around and not attempting to do much apart from that. One interpretation of all of this is that the mice have stopped trying to swim because they are now "depressed" and have given up. Does that seem like a bit of a stretch? To be fair, many researchers now don't believe that tests like the forced swim tests are really good representations of human depression, and the search is on for something better. But as I discuss here, that is probably a fool's errand.

Another variation on this form of testing, which is generally called "learned helplessness," involves just picking up a mouse by its tail and hanging it upside down. Mice eventually give up struggling and trying to right their position and just hang there. Clearly the mouse is depressed! Of course, there are other reasons mice might stop doing these things. For example, they could be tired or simply bored, but such possibilities aren't as interesting to scientists who want to study depression. There are other ways of assessing whether a mouse is depressed by doing something to them like administering unavoidable electric shocks. Are the mice now depressed? Perhaps you will observe that their desire to drink sweetened water is reduced (presumably mouse anhedonia) or that they groom less (presumably mouse feelings of worthlessness). After all, if you did these things to humans, they might well become depressed.

One reason people think that these behaviors in animals represent depression is because drugs that reduce depression in humans, such as SSRIs, also reduce some of these behaviors in mice. OK, so let's assume for a moment that such an argument is valid. What a scientist would now do is try and find new types of drugs that also reduce these mouse behaviors. The argument would be that such drugs would also be antidepressants when given to humans. What kinds of drugs might be tested? Is there any way of getting some guidance on what to test? Nowadays, you might look at large data sets obtained from genetic analysis of depressed human patients and see if there are any genes whose expression seems to be frequently altered in these patients. Perhaps these genes might be involved in producing the depression? Now you "knock out" (remove) these genes from mice using genetic engineering and see if a depression-associated behavior like learned helplessness is increased or reduced. It is! Now all you have to do is make a drug that does the same thing, and you will have a new type of antidepressant. On the other hand, you might select drug targets that are suggested by a particular hypothesis that is already in the scientific literature, for example, that depression is due to stress (the corticotropin-releasing factor [CRF] hypothesis)[56] or abnormalities in the development of nerves (the brain-derived neurotrophic factor [BDNF] hypothesis).[57] These types of approaches have been used millions of times. So, how many new antidepressant drugs have been produced from these studies? The answer, as we discussed in the previous chapter, is that animal studies are rarely effective in this regard, particularly when one considers the enormous number of scientific papers published on this topic. And why is that, given the fact that the entire road map seems quite reasonable from a scientific point of view? The answer, of course, is that what we are measuring in mice isn't a model of human depression. Similar things could be said of other disorders of the human psyche, particularly schizophrenia, as well as other conditions in which human consciousness is of great importance, like pain. Of course, there are drugs for treating human depression or schizophrenia. They are fantastically important, and they completely revolutionized the treatment of these diseases starting in the 1950s and '60s. However, very few of these drugs were discovered by performing animal research but rather through observations on human patients.[58]

Diseases like depression and schizophrenia clearly involve very human aspects of consciousness and self-consciousness in particular. As we descend the evolutionary scale to apes, we might imagine that something akin to human consciousness is still present, although we should also admit that it will not just be a diluted form of human consciousness. Instead, it will have aspects that reflect the needs and physiology of the particular animals in question. This may also be true of animals like cats and dogs, but here the human-like aspects will be even less marked. In a mouse, they will be even less so. The evidence, as we have seen, is that most of these animals have sophisticated conscious experiences, but real mice aren't Beatrix Potter mice, and the actual emotions and "feelings" that they experience will surely not be exactly the same as those of a human. Whatever it may be, it will be unique and appropriate to the animal in question.

Consciousness, as we find it in humans, is presumably something that evolved slowly over many eons.[59] Darwin, in his 1872 monograph *The Expression of the Emotions in Man and Animals*, was the first to consider the way in which this kind of thing may have happened. There will surely be aspects of consciousness in most animals, even perhaps traces in animals like insects. Clever evo/devo biologists and geneticists may be able to mine data from such creatures so that we come to understand how and why consciousness appeared in the first place and how it evolved. If, as with most traits, what we see in humans and animals is the result of an evolutionary history, we would imagine that such a thing is true of consciousness as well. It would seem obvious that consciousness would have evolutionary advantages in terms of the ability of an organism to survive and procreate. But, as with most other things, it is unlikely that consciousness suddenly emerged fully formed. The human situation where the particular phenomenon of self-consciousness appears to have developed to the highest degree is presumably the result of an enormous amount of evolutionary experimentation that produced a phenomenon that reflects uniquely human needs.

How, then, can a nonhuman animal like a mouse be used to model the type of human experience present in a disease like depression or schizophrenia? In these diseases there is no obvious pathology that we can use as a road map for creating an animal model. In cancer there is a tumor; in heart disease an artery is clogged. Even in some diseases that affect human conscious experience such as Parkinson's disease, or Alzheimer's disease there are clear signs of neuropathology that we might model. In diseases like depression that primarily affect our conscious life there is nothing like that. The result is that the animal models that we employ, as discussed above, are useless because they are mere eidolons of the diseases in question. They are behavioral malapropisms. No wonder they haven't been successful in helping us invent new drugs or other treatments. Nor will they ever be. But diseases like depression are huge problems in medicine. They are not only debilitating but also extremely long-lasting. Their toll on the patient, the health care system, and society is considerable, and so billions of dollars have been spent performing experiments on mice in the misguided hope that they will help to solve problems like depression or schizophrenia. And these experiments are cruel! We think we can make animals depressed by drowning them, electrocuting them, starving them, making them fearful, and abusing them in countless ways. Still, they are never going to be depressed in the same way as a human, and they will never be useful in this regard.

Consciousness is clearly connected to the activity of the nervous system in some way. The effects of injury, surgical and electrical interventions, and drugs like hallucinogens suggest this beyond a reasonable doubt, and the scientific evidence would strongly suggest that we can attribute forms of consciousness to animals as well. Although we will presumably never know exactly "what it is like to be a mouse," it is clearly like "something," and this something is reflected by facial expressions, sounds, behaviors, and brain activity that is indicative of some kind of life that is imbued with emotions and conscious experiences that are appropriate for mice. Mice have their own specific internal life and qualia; they are not just different versions of humans. Very few people deny this conclusion anymore. Indeed, on July 7, 2012, a prominent group of scientists signed the Cambridge Declaration on Consciousness.[60] This declaration stated that not only humans but also a significant number of animals, including not just vertebrates but also many invertebrates, are conscious beings. What this means is that they experience what happens to them and have mental states that can be positive or negative for them. And this is an extremely important point. Whatever the precise characteristics of animal consciousness may be, the fact that at some level things are good or bad for them is important. The nineteenth-century English philosopher Jeremy Bentham probably best encapsulated the situation by asking of animals, "The question is not, can they reason? nor, can they talk? but, can they suffer?"[61]

The answer to Bentham's question is clearly yes, they can certainly suffer. It should be clear that research on animals, at least the vast majority of research, which is supposed to translate animal results to humans, is specifically designed to make animals suffer. An aspect of all human disease is human suffering. Otherwise, why would we care? When we use animals as models, it is because we are investigating human disease and therefore, inevitably, human suffering. An animal model of a disease that does not include suffering is not a good animal model. Generally speaking, we do not use animal models to investigate why human beings are happy or content. Pain, cancer, and arthritis are entirely negative experiences for humans. We use animals to model human diseases so that we do not have to make humans suffer in this way. And, of course, we are used to having others suffer on our behalf. It is not surprising that the use of animals to model human diseases arose in Christian Europe under the watchful eye of the church. We have been taught, and this is well ingrained in all our minds, even lurking in the recesses of the brains of scientists who declare themselves convinced atheists, that we are the pinnacle of creation, made in the image of God. Christ suffered for us, so why wouldn't it be natural to allow animals to do so as well? Surely anything can be sacrificed on the altar of human self-regard. However, all this comes at a terrible cost. When we brutalize other humans or animals that can clearly suffer, we degrade ourselves and the core of our humanity. We may be proud of our achievements intellectually, but still, they shame us. We may rule Earth, but we do so through the use and brutal exploitation of other creatures. Like Ozymandias, we may be the king of kings, but we should look upon our works and despair.

Notes

1. Ludwig Wittgenstein et al., *Philosophische Untersuchungen* [Philosophical investigations], rev. 4th ed. (Chichester, West Sussex, UK; Malden, MA: Wiley-Blackwell, 2009).

2. Thomas Browne and C. A. Patrides, *The Major Works*, The Penguin English Library (Harmondsworth, UK; New York: Penguin Books, 1977), chap. 5, Hydriotaphia: Urne burial, 307.

3. There are, of course, innumerable stories of animals with human characteristics including, for example, "Mrs. Frisby and the rats of NIMH," which tells the story of laboratory rats who have been made super-intelligent by scientists at the National Institute of Mental Health. Indeed, they are now smart enough to escape and seek revenge on their scientific oppressors. Mouse genetic engineers might want to think about that one. See Robert C. O'Brien and Zena Bernstein, *Mrs. Frisby and the Rats of Nimh* (New York: Aladdin Books, 1986).

4. Susan Orlean, *On Animals* (S.l.: Avid Reader PR, 2022), 5.

5. Nate Dolensek et al., "Facial Expressions of Emotion States and Their Neuronal Correlates in Mice," *Science* 368, no. 6486 (April 2020): 89–94, doi:10.1126/science.aaz9468.

6. Daniel T. Sangiamo, Megan R. Warren, and Joshua P. Neunuebel, "Ultrasonic Signals Associated with Different Types of Social Behavior of Mice," *Nature Neuroscience* 23, no. 3 (March 2020): 411–422, doi:10.1038/s41593-020-0584-z.

7. Recent studies have also shown that individual colonies of some rodents (mole rats) even have their own dialects—Rochelle Buffenstein, "Colony-Specific Dialects of Naked Mole-Rats," *Science* 371, no. 6528 (January 2021): 461–462, doi:10.1126/science.abf7962.

8. Annika Stefanie Reinhold et al., "Behavioral and Neural Correlates of Hide-and-Seek in Rats," *Science* 365, no. 6458 (September 2019): 1180–1183, doi:10.1126/science.aax4705.

9. S. Ishiyama and M. Brecht, "Neural Correlates of Ticklishness in the Rat Somatosensory Cortex," *Science* 354, no. 6313 (November 2016): 757–760, doi:10.1126/science.aah5114.

10. Emily Underwood, "Rats Forsake Chocolate to Save a Drowning Companion," *Science*, May 12, 2015, doi:10.1126/science.aac4586.

11. Monique L. Smith et al., "Social Transfer of Pain in Mice," *Science Advances* 2, no. 10 (October 2016): e1600855, doi:10.1126/sciadv.1600855; Monique L. Smith, Naoyuki Asada, and Robert C. Malenka, "Anterior Cingulate Inputs to Nucleus Accumbens Control the Social Transfer of Pain and Analgesia," *Science* 371, no. 6525 (January 2021): 153–159, doi:10.1126/science.abe3040.

12. Frans B. M. de Waal and Stephanie D. Preston, "Mammalian Empathy: Behavioural Manifestations and Neural Basis," *Nature Reviews Neuroscience* 18, no. 8 (August 2017): 498–509, doi:10.1038/nrn.2017.72; F. B. M. de Waal, *Mama's Last Hug: Animal Emotions and What They Tell Us about Ourselves* (New York: W. W. Norton & Company, 2019).

13. Sarah Cascone, "Bored in Self-Isolation, an Art-World Power Couple Built a Tiny Museum for Their Gerbils, Complete with Rodent Leonardos and Vermeers," *Art World*, April 2020, https://news.artnet.com/art-world/gerbil-art-museum-london-1827057.

14. Richard Kearney and David M. Rasmussen, eds., *Continental Aesthetics: Romanticism to Postmodernism: An Anthology*, Blackwell Philosophy Anthologies 12 (Malden, MA: Blackwell Publishers, 2001).

15. Eric Edsinger and Gül Dölen, "A Conserved Role for Serotonergic Neurotransmission in Mediating Social Behavior in Octopus," *Current Biology* 28, no. 19 (October 2018): 3136–3142.e4, doi:10.1016/j.cub.2018.07.061.

16. Edsinger and Dölen, "A Conserved Role for Serotonergic Neurotransmission in Mediating Social Behavior in Octopus."

17. Alexandra K. Schnell et al., "Cuttlefish Exert Self-Control in a Delay of Gratification Task," *Proceedings of the Royal Society B: Biological Sciences* 288, no. 1946 (March 2021): 20203161, doi:10.1098/rspb.2020.3161.

18. Dominique Potvin, "Altruism in Birds? Magpies Have Outwitted Scientists by Helping Each Other Remove Tracking Devices," *The Scientist*, February 2022, https://www.the-scientist.com/news-opinion/altruism-in-birds-magpies-have-outwitted-scientists-by-helping-each-other-remove-tracking-devices-69723.

19. Suzana Herculano-Houzel, "Birds Do Have a Brain Cortex—and Think," *Science* 369, no. 6511 (September 2020): 1567–1568, doi:10.1126/science.abe0536.

20. Elizabeth Pennisi, "2020 Breakthrough of the Year," *Science*, December 17, 2020, https://vis.sciencemag.org/breakthrough2020/#/finalists/birds-are-smarter-than-you-think.

21. "Crows Perform yet Another Skill Once Thought Distinctively Human," https://www.scientificamerican.com/article/crows-perform-yet-another-skill-once-thought-distinctively-human/.

22. Ludwig Wittgenstein, *Philosophical Investigations* (Oxford: Basil Blackwell, 1968).

23. Tod Machover, "Death and the Powers," 2012, https://opera.media.mit.edu/projects/deathandthepowers/.

24. Emily Underwood, "A Sense of Self," *Science* 372, no. 6547 (June 2021): 1142–1145, doi:10.1126/science.372.6547.1142.

25. Thomas Nagel, "What Is It Like to Be a Bat?," *Philosophical Review* 83, no. 4 (October 1974): 435, doi:10.2307/2183914.

26. Natasha Frost, "What Is It Like to Be a Bee?," *Atlas Obscura*, December 2017.

27. Henry Beston, *The Outermost House: A Year of Life on the Great Beach of Cape Cod* (New York: Holt paperbacks, 2003).

28. Sara Asu Schroer, "Jakob von Uexküll: The Concept of *Umwelt* and Its Potentials for an Anthropology Beyond the Human," *Ethnos* 86, no. 1 (January 2021): 132–152, doi:10.1080/00141844.2019.1606841.

29. Catherine Offord, "A Dog's View of Optical Illusions," *The Scientist*, January 2021, https://www.the-scientist.com/features/a-dogs-view-of-optical-illusions-68278.

30. Torin Alter and Sven Walter, *Phenomenal Concepts and Phenomenal Knowledge* (Oxford: Oxford University Press, 2007), doi:10.1093/acprof:oso/9780195171655.001.0001.

31. Stanislas Dehaene, *Consciousness and the Brain: Deciphering How the Brain Codes Our Thoughts* (New York: Penguin Books, 2014).

32. Joseph E. LeDoux and Caio Sorrentino, *The Deep History of Ourselves: The Four-Billion-Year Story of How We Got Conscious Brains* (New York: Viking, 2019); Susan J. Blackmore, *Consciousness: A Very Short Introduction*, Very Short Introductions (Oxford, UK; New York: Oxford University Press, 2005); Lukas J. Volz and Michael S. Gazzaniga, "Interaction in Isolation: 50 Years of Insights from Split-Brain Research," *Brain* 140, no. 7 (July 2017): 2051–2060, doi:10.1093/brain/awx139.

33. Jonathan Swift and Albert J. Rivero, *Gulliver's Travels: Based on the 1726 Text: Contexts, Criticism*, A Norton Critical Edition (New York: Norton, 2002).

The results of "split brain" experiments had been considered, at least theoretically, hundreds of years ago by Jonathan Swift, who had Lemuel Gulliver comment on them when visiting a virtuoso (scientist) living in the flying island of Laputa: "When parties in a state are violent, he offered a wonderful contrivance to reconcile them.

The method is this. You take a hundred leaders of each party, you dispose them into couples of such whose heads are nearest of a size; then let two nice operators saw off the occiput of each couple at the same time, in such a manner that the brain may be equally divided. Let the occiputs this cut-off be interchanged applying each to the head of the opposite Party-man. It seems indeed to be a work that requireth some exactness, but the professor assured us, that if it were dexterously performed, the cure would be infallible. For he argued thus; that the two half brains being left to debate the matter between themselves with the space of one scull, would soon come to a good understanding, and produce that moderation as well as regularity of thinking, so much to be wished for in the heads of those, who imagine they came into the world only to watch and govern its motion; and as to the difference of brains in quantity of quality, among those who are directors in faction: the doctor assured us from his own knowledge, that it was a perfect trifle."

34. Sam Kean, "This Man Can Read Letters but Numbers Are a Blank," *Science* 369, no. 6503 (July 2020): 494–494, doi:10.1126/science.369.6503.494.

35. See discussion in Richard J. Miller, *Drugged: The Science and Culture behind Psychotropic Drugs* (Oxford: Oxford University Press, 2014).

36. Stanislav Grof and Albert Hofmann, *LSD Psychotherapy: The Healing Potential of Psychedelic Medicine*, 4th ed (Ben Lomond, CA: MAPS, 2008).

37. Grof and Hofmann, *LSD Psychotherapy*.

38. Miller, *Drugged*.

39. David P. Luke and Devin B. Terhune, "The Induction of Synaesthesia with Chemical Agents: A Systematic Review," *Frontiers in Psychology* 4, no. 1–12 (October 17, 2013), doi:10.3389/fpsyg.2013.00753.

40. Marcus E. Raichle, "The Brain's Default Mode Network," *Annual Review of Neuroscience* 38, no. 1 (July 2015): 433–447, doi:10.1146/annurev-neuro-071013-014030.

41. Enzo Tagliazucchi et al., "Increased Global Functional Connectivity Correlates with LSD-Induced Ego Dissolution," *Current Biology* 26, no. 8 (April 2016): 1043–1050, doi:10.1016/j.cub.2016.02.010.

    In a useful summary of their results in scientific language, the authors explain, "Here we studied the effects of LSD on intrinsic functional connectivity within the human brain using fMRI. High-level association cortices (partially overlapping with the default-mode, salience, and frontoparietal attention networks) and the thalamus showed increased global connectivity under the drug. The cortical areas showing increased global connectivity overlapped significantly with a map of serotonin 2A (5-HT2A) receptor densities (the key site of action of psychedelic drugs). LSD also increased global integration by inflating the level of communication between normally distinct brain networks. The increase in global connectivity observed under LSD correlated with subjective reports of 'ego-dissolution.'"

42. Tagliazucchi et al., "Increased Global Functional Connectivity Correlates with LSD-Induced Ego Dissolution."

43. Gregory Scott and Robin L. Carhart-Harris, "Psychedelics as a Treatment for Disorders of Consciousness," *Neuroscience of Consciousness* 5, no. 1 (January 2019): 1–8, doi:10.1093/nc/niz003.

44. Robin L. Carhart-Harris et al., "The Entropic Brain: A Theory of Conscious States Informed by Neuroimaging Research with Psychedelic Drugs," *Frontiers in Human Neuroscience* 8 (2014): 1–22, doi:10.3389/fnhum.2014.00020.

45. Alison Abbott, "Inside the Mind of an Animal," *Nature* 584, no. 7820 (August 2020): 182–185, doi:10.1038/d41586-020-02337-x.

  Brain imaging measurements do not actually represent a picture of thinking and it is all too easy to anthropomorphize them and fall into the trap that we are observing thoughts. In reality, these pictures are not thoughts; they are just another set of measurements that seem to correlate with conscious experience in interesting ways. Nevertheless, they do introduce a different set of terms into the conversation and what is clearly a valuable perspective. Live imaging experiments, then, provide us with another experimental perspective on what is going on in the brain and how this might be related to different "types" of consciousness. A recent article in the journal *Nature* was entitled "What Animals Really Think" but, disappointingly, just described the latest set of ever more complex electrophysiological and imaging measurements. No animal thoughts were reported on—alas!

46. D. Mantini et al., "Default Mode of Brain Function in Monkeys," *Journal of Neuroscience* 31, no. 36 (September 2011): 12954–12962, doi:10.1523/JNEUROSCI.2318-11.2011.

47. James M. Stafford et al., "Large-Scale Topology and the Default Mode Network in the Mouse Connectome," *Proceedings of the National Academy of Sciences* 111, no. 52 (December 2014): 18745–18750, doi:10.1073/pnas.1404346111.

48. Paca Thomas and Jeffrey Bloomer, "The Strange, Sordid History of Dosing Animals With LSD," *SLATE*, July 2014, https://slate.com/technology/2014/07/animal-lsd-experiments-elephants-dolphins-fish-given-hallucinogenic-drugs-video.html.

49. Adam L. Halberstadt et al., "Correlation between the Potency of Hallucinogens in the Mouse Head-Twitch Response Assay and Their Behavioral and Subjective Effects in Other Species," *Neuropharmacology* 167 (May 2020): 107933, doi:10.1016/j.neuropharm.2019.107933.

50. Angie M. Michaiel, Philip R. L. Parker, and Cristopher M. Niell, "A Hallucinogenic Serotonin-2A Receptor Agonist Reduces Visual Response Gain and Alters Temporal Dynamics in Mouse V1," *Cell Reports* 26, no. 13 (March 2019): 3475–3483.e4, doi:10.1016/j.celrep.2019.02.104.

51. "Mirror Test," *Wikipedia*, n.d., accessed July 13, 2022, https://en.wikipedia.org/wiki/Mirror_test; Carolyn Wilke Wilke, "The Mirror Test Peers Into the Workings of Animal Minds," *The Scientist*, February 2019, https://www.the-scientist.com/news-opinion/the-mirror-test-peers-into-the-workings-of-animal-minds-65497.

52. Lawrence Wright, "The Elephant in the Courtroom," *New Yorker*, February 2022, https://www.newyorker.com/magazine/2022/03/07/the-elephant-in-the-courtroom.

53. Erin McCallum, "Are Fish Self-Aware?," *Journal of Experimental Biology* 222, no. 9 (May 2019): jeb192781, doi:10.1242/jeb.192781.

54. Marie-Claire Cammaerts and Roger Cammaerts, "Are Ants (Hymenoptera, Formicidae) Capable of Self Recognition?," *Journal of Science* 5, no. 7 (2015): 521–532, http://www.journalofscience.net/showpdf/MjY4a2FsYWkxNDc4NTIzNjk=.

55. Barbara Planchez, Alexandre Surget, and Catherine Belzung, "Animal Models of Major Depression: Drawbacks and Challenges," *Journal of Neural Transmission* 126, no. 11 (November 2019): 1383–1408, doi:10.1007/s00702-019-02084-y.

56. Elena Goetz Davis et al., "Corticotropin-Releasing Factor 1 Receptor Haplotype and Cognitive Features of Major Depression," *Translational Psychiatry* 8, no. 1 (December 2018): 5, doi:10.1038/s41398-017-0051-0.

57. Chin-Chuen Lin and Tiao-Lai Huang, "Brain-Derived Neurotrophic Factor and Mental Disorders," *Biomedical Journal* 43, no. 2 (April 2020): 134–142, doi:10.1016/j.bj.2020.01.001.

58. See Miller, *Drugged*, 203–235, for an exception. The class of drugs known as anxiolytics or antianxiety drugs, which includes things like meprobamate or the benzodiazepines, was discovered by observing their effects on animal behavior by the Hofmann-LaRoche drug company laboratories in Nutley, New Jersey, in the 1960s prior to any indications of their activity in humans. Of course, having discovered these types of drugs, the discovery of subsequent iterations of the initial exemplars of each class has certainly been aided by using animals, but no new drug classes have been discovered. Many drug companies have now become so disillusioned with the entire enterprise that they have stopped doing animal-based research in these areas.

59. David J. Anderson and Ralph Adolphs, "A Framework for Studying Emotions across Species," *Cell* 157, no. 1 (March 2014): 187–200, doi:10.1016/j.cell.2014.03.003.

60. Marc Bekoff, "Animals Are Conscious and Should Be Treated as Such," *New Scientist*, September 2012, https://www.newscientist.com/article/mg21528836-200-animals-are-conscious-and-should-be-treated-as-such/.

61. Jeremy Bentham et al., *An Introduction to the Principles of Morals and Legislation*, The Collected Works of Jeremy Bentham (Oxford; New York: Clarendon Press; Oxford University Press, 1996).

    "Other animals, which, on account of their interests having been neglected by the insensibility of the ancient jurists, stand degraded into the class of things. . . . The day has been, I grieve it to say in many places it is not yet past, in which the greater part of the species, under the denomination of slaves, have been treated . . . upon the same footing as . . . animals are still. The day may come, when the rest of the animal creation may acquire those rights which never could have been withholden from them but by the hand of tyranny. The French have already discovered that the blackness of skin is no reason why a human being should be abandoned without redress to the caprice of a tormentor. It may come one day to be recognized, that the number of legs, the villosity of the skin, or the termination of the os sacrum, are reasons equally insufficient for abandoning a sensitive being to the same fate. What else is it that should trace the insuperable line? Is it the faculty of reason, or perhaps, the faculty for discourse? . . . [T]he question is not, Can they reason? nor, Can they talk? but, Can they suffer? Why should the law refuse its protection to any sensitive being? . . . The time will come when humanity will extend its mantle over everything which breathes."

6

The Modern Prometheus

Death, thou shalt die!

—Holy Sonnet No. 10, John Donne

Oh Man! Listen carefully!
What does the deep midnight say?
I slept, I slept -
And from deepest dreams have I awakened.

So begins the "Midnight Song" or "Zarathustra's Roundelay" from Friedrich Nietzsche's 1885 book *Thus Spoke Zarathustra*. Surely this is a description of the transition between life and death? We sleep. The molecules of our being are scattered throughout eternity, and then something miraculous occurs. Energy provided by the sun allows these particles to coalesce to produce life and so, "from deepest dreams have we awakened." We have defied entropy and live far from thermodynamic equilibrium for the number of our days, until eventually we return once again to particles of matter distributed to the ends of time, back once more to our original slumber. Life, then, is that brief instant when we take on a particular material form. But even today exactly what it means to be alive is something that is hard to define. The material that makes up a living being, as opposed to a corpse, is not so different and yet something is missing. As a result, we perpetually ask ourselves the same questions: Is the transition between life and death really irreversible? How can we produce life from an inanimate object? What is it that we need to add? What is the secret? Answering questions such as these was something that was considered to be of considerable importance by many ancient sages. Writings in the Hermetic tradition tell us clearly how to conjure up spirits and draw them down from the macrocosm into idols so they can live and prophecy. Hermes Trismegistus revealed these secrets to his son Asclepius. Great thinkers in Renaissance Italy like Marsilio Ficino and Giovanni Pico de

The Rise and Fall of Animal Experimentation. Richard J. Miller, Oxford University Press. © Richard J. Miller 2023.
DOI: 10.1093/oso/9780197665756.003.0006

Mirandola were extremely impressed when they discovered these revelations and sought to incorporate them into Christian thinking. Are such acts of animation possible? What is it that we can give back to a dead body that would make it live again? Is there an *élan vital,* something that is beyond the normal world of the organic and inorganic, as vitalists down through the centuries would have us believe?

Or is it, as hinted at by people like Galvani and made clear in Mary Shelley's *Frankenstein,* something we can appropriate from the world of science, like electricity and magnetism? In a prologue to the 1831 edition of *Frankenstein,* Mary Shelley described the influence bioelectricity (galvanism) had on her thinking, in this instance due to an observation made by Erasmus Darwin, Charles Darwin's grandfather.

> Many and long were the conversations between Lord Byron and Shelley, to which I was a devout but nearly silent listener. During one of these, various philosophical doctrines were discussed, and among others the nature of the principle of life, and whether there was any probability of its ever being discovered and communicated. They talked of the experiments of Dr. Darwin, who preserved a piece of vermicelli in a glass case, till by some extraordinary means it began to move with voluntary motion. Not thus, after all, would life be given. Perhaps a corpse would be re-animated; galvanism had given token of such things: perhaps the component parts of a creature might be manufactured, brought together, and endued with vital warmth.[1]

Nowadays, a unique vital spirit seems like a simplistic idea; life is not just about one thing like adding yeast to bread and getting it to rise. We would certainly say from our modern perspective that life isn't a single process but a set of interlocking systems of great complexity. But understanding exactly how this works in practice has eluded us.

Of course, our general ignorance as to what precisely contributes to and sustains life has not stopped us speculating about it, attempting to produce it, and, from the very beginning of our efforts, worrying about the potential consequences of doing so. Writing at a time when the winds of change were carrying the concepts of Romantic biology throughout the world, Mary Shelley warned us about the possible insanity that might result from our attempts, and her message has been often repeated in stories like H. P. Lovecraft's "Herbert West—Reanimator" down to the present day. For Herbert West, the secret wasn't electricity, but a powerful fluid extracted from freshly disinterred corpses that could be injected into the dead. They came back to life but were horribly changed.

Even the most up-to-date thinking in the sciences and the humanities struggles to define what death actually entails and how irreversible a state it really is. In most countries, a person can be legally declared dead if they show irreversible loss of all brain functions (brain death) or irreversible loss of all circulatory cardiovascular functions (circulatory death). This implies, of course, that such things cannot be reliably restored. Most people would probably agree that this seems reasonable; if the brain and heart show no activity, then what remains?

Surprisingly, the contemporary scientific answer to this question would be "quite a lot." The brain, for example, the most fragile of all our organs, requires a great deal of energy to work properly, and even a few minutes without oxygen, something that might occur during a stroke, usually leaves the brain "irreversibly" damaged, producing lifelong problems with movement and cognition. Indeed, a serious stroke is frequently fatal. In such cases the brain ceases to function and that is that; there is no going back. That is what most of us believe. *Frankenstein* is a nice story, but it's just a story. In 2019, this conclusion was tested by a group of researchers at Yale University and published in the journal *Nature*.[1] They took the brains from 300 pigs that had been slaughtered and waited four hours. They then hooked the brains up to a perfusion system containing a nutrient-rich solution called BrainEx. The result of this procedure was remarkable. The investigators observed that many brain functions were restored. Blood vessels in the brain could be made to dilate and contract like a normal brain and so on. When nerve cells in slices of the treated brains were examined in a laboratory, they were able to fire action potentials, the electrical signals that are fundamental to a functioning brain. On the other hand, brains that hadn't been perfused with Brainex showed no activity of any kind. Subsequently, in 2022, experiments with a similar solution called OrganEx revived activity of other organs throughout the body. The implications of these observations for fantasies like *Frankenstein* and Herbert West are clear enough; dead is only dead if science hasn't advanced enough to address serious biological problems that normally result in the cessation of key physiological functions. And this changes all the time. Somebody who died from a heart attack in the 1950s may well not suffer the same fate today. Many ethical and other issues surrounding scientific advances such as brain revival have only just begun to be discussed, and the answers are unclear. But, be assured, our ability to do this kind of thing is only going to improve, and probably much more quickly than we expect. And if we have a revived human brain at our disposal, what will we do with it? Restore individuals to health, provide tissues for organ transplants-or what?

If we want to carry out research to solve scientific problems that pertain to humans, it would certainly be best if the scientists of the future performed these studies using human tissues. Performing experiments on animals is frequently cruel, unethical, and not particularly helpful if relevance to humans is our goal. Animals are only supposed to act as stand-ins for humans anyway. But if we aren't going to use them, then what? Where exactly are the human tissues that we want to use for our experiments supposed to come from? Should we go back to the Alexandria of the third century BCE, to the days of Herophilus and Erasistratus (see Chapter 2), and resume the practice of human vivisection once more? Such an idea would surely find little support—thank goodness! Moreover, if we are considering performing experiments on humans rather than animals, haven't we just swapped one set of ethical issues for another? However, as things turn out, advances in the biological sciences have started to suggest a way forward so that human experimentation is now possible from both scientific and ethical perspectives, making it likely that animal experimentation will be completely unnecessary in the near future.

Our assumption that the processes leading from life to death are a one-way street is at the very basis of everything we do. The entire idea of medicine is devoted to maintaining our health and extending our lives. True immortality has its issues, like the inevitability of eventually losing interest in everything, as Borges illustrated in his classic story, "The Immortal." Yet most people would agree that they would like to live longer, disease free—at least up to a point.

From a purely scientific point of view, life is just another set of linked events that describe the development of certain kinds of cells through time. We begin with a fertilized egg; the egg develops into an animal, at least long enough for it to reproduce and pass along its genes; and then the animal gradually begins to become less functional until the arrival of death when the lights are switched off. If you are a scientist, you would look at this process and say it's just another problem in cell biology and biochemistry for us to understand. It's a process that plays by certain rules that we can, at least in principle, decipher, and why shouldn't we? Once stripped of its metaphysical trappings, life doesn't seem that special after all.

What, then, is an irreversible process in biology? If a dead brain can be reanimated, perhaps there are other similar phenomena? Reviving the activity of a brain is a functional intervention. Are there also material interventions? If tissues are destroyed, can they be regrown and replaced?

The Birth of Regeneration

There are certainly examples of myths of regeneration, even in antiquity. Indeed, as described in Hesiod's *Theogony*, one might consider the Titan Prometheus, who stole fire from the gods and gave it to humans, a race of beings he had created from clay. Zeus punished Prometheus by chaining him to a rock on a mountain in the Caucasus. Every day an eagle came and devoured his liver. Then, each night the liver regenerated itself, and the next day the eagle returned. This was supposed to go on forever, and in the myth, it lasted for thirteen generations. Then, Heracles came to Prometheus's rescue and killed the eagle. What prompted the development of a legend such as this in antiquity? The fact is that in many animals, a damaged liver can regenerate itself. In biomedical research the classical model of hepatic regeneration is what is known as partial hepatectomy in which perhaps 70% of the liver is removed. But in a rat, a week later the liver has regrown, something that has also been demonstrated to occur in humans. It's quite remarkable. The liver regrows to precisely the right size—not too big and not too small.[2]

It is clearly the case that in ancient times, the phenomenon of liver regeneration had found its way into myth. And it wasn't just the regeneration of the liver. Some animals can regenerate an entire limb or other body part. In ancient Greece Aristotle had commented on this phenomenon: "The tails of lizard and of serpents, if they be cut off, will grow again"[3] With the advent of modern experimental science in the sixteenth and seventeenth centuries, observations like these began to take on renewed significance as scientists became eager to attack them experimentally and provide a mechanism for understanding them. As is often the case, new technology enabled scientific progress. It wasn't until the seventeenth century that scientists like Van Leeuwenhoek and Hooke developed the use of the microscope and gave scientists the ability to observe the structure of animals at the level of single cells. This was a key breakthrough. It wasn't long before the microscope became an essential part of every biologist's repertoire of techniques. If you found something new, one of the first things you did was to look at its microscopic structure. Consider the eighteenth-century biologist Abraham Trembley.[4] Trembley was a native of Geneva and had taken a job as tutor to the children of Count William Bentinck in the Netherlands.[5] One day, he was looking under the microscope at samples of plants he had taken from the river, and he observed some tiny organisms that he couldn't identify. They seemed like little tubes that had a head with little arms and feet. He wasn't quite sure if they were plants or animals. He began to experiment with them, chopping them into pieces to see what would happen. To Tremblay's surprise, each tiny piece grew into an entire animal—head, arms, body, and feet. In another experiment, he split an animal from its head to its middle. Amazingly, the two branches both grew new heads. Tremblay thought it resembled the many-headed serpent from Greek and Roman mythology and named the creature hydra. In fact, an entire hydra can be recovered from a segment composed of 1% of the original organism's total volume, a seemingly prodigious feat of regeneration.

Extensive regeneration can not only be observed in hydra.[6] Consider animals such as urodeles (newts and salamanders). Though urodele regeneration had been observed in antiquity, it became a matter of scientific interest in the mid-1760s when Lazzaro Spallanzani (1729–1799), a professor of philosophy at the university in Modena in Italy, discovered that newts and salamanders whose tails and limbs had been removed could regenerate them. Indeed, animals like axolotls (Mexican salamanders) exhibit a capacity for regeneration that is extremely significant for a vertebrate. Humans, on the other hand, seem to have lost this capacity to a great extent, although we can regenerate the tips of our fingers and toes as well as some tissues such as the liver, particularly as young children.[7] In other words, animals can regenerate certain tissues; it just depends on the animal and the tissue under consideration. This raises the question, how do these things happen, and what are the limits to the regeneration of human tissues?

Our modern understanding of these processes really began in the laboratory of Johannes Muller in Berlin in the first half of the nineteenth century. Muller's laboratory hosted a number of outstanding scientists who made seminal contributions to biology.[8] These included the likes of Theodor Schwann, Robert Remak, and Rudolf Virchow, who contributed to the science of cell biology in particular. Although the term "cell" had been coined by Robert Hooke in the seventeenth century, a full understanding of the role of cells in biology was slow in coming. It was Schwann whose studies allowed him to conclude, "All living things are composed of cells and cell products." Building on this principle, his colleague Rudolf Virchow was one of the originators of the idea that cells arose from the division of other cells— *Omnis cellula e cellula* ("all cells come from cells"). Hence, a discussion of the properties of the new science of cells was part of the culture of the laboratory, and it was taken up and developed in different ways.

Of particular importance for the present discussion was the work of Ernst Haeckel,[9] who joined the laboratory and worked with Virchow among others. He became fully conversant with the developing concept of the cell as it was understood at the time. Subsequently, Haeckel joined the faculty of the University of Jena where, in 1861, his colleague Carl Gegenbaur recognized that all eggs were "egg-cells," and hence, that cells constituted the beginning of all embryological development. Haeckel, as we have already discussed (see Chapter 4), was interested in evolutionary theory and became a friend and disciple of Charles Darwin. Haeckel was also a wonderful artist and often thought about science in particularly visual terms. It was Haeckel who first displayed Darwin's ideas about evolution as a picture of a tree, something he called a *Stammbaum* (stem tree) (Figure 6.1), allowing visualization of the relationship of one species to another.[10]

Haeckel developed the notion that ontogeny (embryological development) could be thought of in the same way as phylogeny (evolutionary development). Haeckel's "biogenetic law," which posited that ontogeny recapitulated phylogeny, prompted him to use the term "stem-cell" (*Stammzelle)* to describe the fertilized egg from which an animal rather than a species developed.

> I have given a special name to the new cell from which the child develops, and which is generally loosely called "the fertilized ovum" or "the first seg-mentation sphere." I call it "the stem-cell." . . . The name stem-cell seems to me the simplest and most suitable because all the other cells of the body are derived from it, and because it is, in the strictest sense, the stem-father and stem-mother of all the countless generations of cells of which the multicel-lular organism is to be composed.

In this respect, Haeckel thought that a stem cell was different from a highly differentiated cell like a neuron, which had limited developmental potential, writing in 1868:

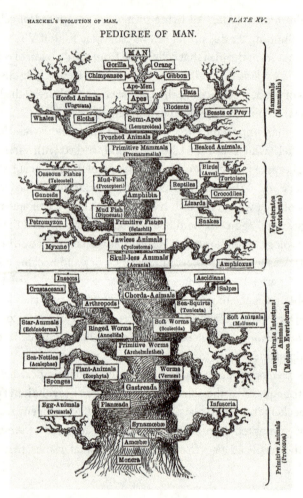

Figure 6.1 Haeckel's Stammbaum.
Image from Alamy.

The ovum stands potentially for the entire organism—in other words, it has the faculty of building up out of itself the whole multicellular body. It is the common parent of all the countless generations of cells which form the different tissues of the body; it unites all their powers in itself, though only potentially or in the germ. In complete contrast to this, the neural cell in the brain, develops along one rigid line. It cannot, like the ovum, beget endless generations of cells, of which some will become skin cells, others muscle cells and others again bone cells.

What Haeckel had really put his finger on was a principle that is key to understanding the development of all animals—the principle of regeneration and perhaps even immortality. It's the idea that a tissue is capable of being not only generated but also regenerated. Haeckel concluded that the fertilized egg, or *zygote*, had the properties of a "totipotent" stem cell, containing all the maternal and paternal influences for directing the subsequent development of a complete organism.

Haeckel attempted to provide experimental evidence for his ideas as Tremblay had done before him. He traveled to Lanzarote in the Canary Islands and began to investigate the sea creatures in the area. He was immediately struck by the ethereal beauty of the *Cnidaria*, a group of ancient animals that includes jellyfish.[11] His attention was particularly drawn to creatures named siphonophores that are closely related to the hydra that Abraham Trembley used in his investigations (Figure 6.2). Haeckel had developed a method whereby he could use one eye to look down the microscope while his other eye and hand produced drawings of what he saw. He made some truly wonderful drawings of these creatures and began to investigate their structure and habits. Indeed, Haeckel's swirling images had a considerable influence on the development of the late nineteenth-century art form known as Art Nouveau, Jugendstil, or Stile Liberty.[12]

One group of experiments resembled Tremblay's studies of hydra regeneration but taken to a more cellular level. Haeckel took two-day-old siphonophore embryos and, using a fine needle and microscope, divided them into small groups of cells. He then placed each cell or cell group into a dish of seawater and observed what happened to them over the next few days. To his surprise, in numerous cases, these cells or cellular fragments regenerated entire intact embryos. Clearly, the cells of these early embryos were capable of regenerating the entire organism—they were indeed stem cells. Haeckel's concept of the stem cell was something that many scientists found useful in their thinking about how organisms developed, and it was rapidly incorporated into the language of scientists working all over the world. For example, histopathologists subsequently applied the stem cell concept to normal and leukemic hematopoiesis, putting forward the idea that there was a common progenitor of red and white blood cells as well as a common precursor of myeloid and lymphoid leukemic cells. From the very beginning, the stem cell concept has been imagined as a tree-like process, in which multipotent stem cells give rise to their progeny through an ordered series of branching steps.

Figure 6.2 Siphonophore from Haeckel's *Art Forms in Nature*.
Image from Alamy.

The most widely debated theory of embryological development during the late nineteenth century had been suggested by the great developmental biologist August Weismann, an enormously influential figure.[13] Weismann used the term "Germplasm" (*Keimplasma*) to describe the material, already known at this time to be associated with the nucleus of the cell, that carried genetic information from one generation to another. Weismann supposed that there were two types of cells: the germ cells, sperm and egg cells, that retained the entire germplasm and so passed it on from one generation to the next, and the somatic cells that made up the different tissues of the organism (see Figure 6.3). Weismann thought that once the zygote began to divide, the germplasm became "segregated" into different cells. Only the original fertilized egg, the zygote, and its germ cell progeny (sperm and egg cells) would contain all the requisite information for producing an entire organism. On the other hand, cells that were destined to become the different tissues of an organism would have reduced amounts of germplasm—what Weismann described as "somatoplasm." In other words, because these somatic cells only contained reduced amounts of hereditary material, they could only make specific tissues rather than entire organisms. Weismann considered the segregation of hereditary material into the various somatic cell lineages to be an irreversible process. There would be no way that changes to somatic cells could influence the germ line. Hence, he concluded, in contrast to figures such as Lamarck and Darwin, that it would not be possible for acquired traits to influence heredity, a highly influential idea that has shaped our thinking ever since.

The truth of Weismann's supposition was tested by one of Haeckel's students, Wilhelm Roux. Roux took frog embryos at the two-cell stage, killed one cell with a hot needle, and examined what happened to the development of the remaining cell. He observed, presumably in keeping with Weismann's ideas, that the remaining cell only produced half an embryo. However, that was not the end of the story. In 1890, inspired by his colleague's

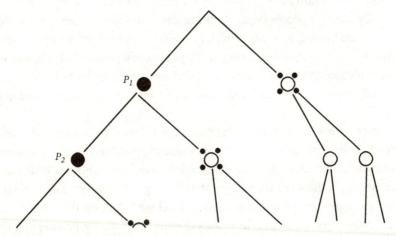

Figure 6.3 Boveri's model of stem cell development. P, developing germ cell.
Image from Alamy.

results, another of Haeckel's students, Hans Driesch, performed a similar set of experiments using sea urchin embryos, an experimental preparation that had been introduced by two more of Haeckel's students, Richard and Oscar Hertwig. He separated two-cell embryos and placed the single cells into separate dishes. Surprisingly, he observed that they all developed into complete sea urchin embryos—exactly the opposite result from Wilhelm Roux's observations using frog embryos. The reason for this difference was unclear at the time, and a new debate was sparked in biology—the Roux-Driesch rivalry.[14] Ultimately, however, attempts to repeat Roux's results failed, and it was subsequently shown that even cells taken from early frog embryos could produce entire frogs.

Further progress on the nature of stem cells also took place in Germany, particularly due to the efforts of Theodor Boveri, who was himself student of Richard Hertwig in Munich. Boveri studied the embryo of the roundworm of the horse (*Ascaris megalocephala*), then a common model organism for cytological research that had been introduced by his teacher Hertwig.

Boveri's observations resulted in a model that suggested that each time a stem cell divided, it produced another stem cell and a second cell that would act as a precursor to some type of somatic cell (see Figure 6.3). In other words, once the early phase of embryological development was completed, one "stem cell" divided into two daughter cells, of which only one maintained the character of a stem cell whereas the other divided into the precursor of the somatic cells that were to become the various tissues. This model found its way into important textbooks in America and Europe and became essentially the way people began to think about the role of stem cells in development. Indeed, it is this model, in which stem cells utilize a process of "asymmetric" cell division during development, that is essentially the way we still think about this process. The properties of stem cells include a capacity for self-renewal as well as for differentiation into specific types of somatic cells.[15]

Nevertheless, at the turn of the twentieth century there were still many things about stem cells that seemed unresolved, particularly with respect to the molecular mechanisms that governed stem cell development. Weismann had proposed that the hereditary material was gradually segregated into the different types of somatic cells as they developed. But, on the other hand, Driesch's observations and the results of cytologists demonstrating that the chromosomal content of cells was preserved during mitotic cell division opposed this idea. Significant answers to some of these questions resulted from the work of yet another German scientist beginning in 1902. This was Hans Spemann, whose model organism of choice was the salamander. He conducted studies a la Roux and Driesch, in which he attempted to separate early salamander embryos into single cells. These embryos were sticky and couldn't just be separated by shaking, the process that had been used for the sea urchin. Instead, Spemann had to fashion a tiny lasso out of a baby's hair and used it to separate salamander embryos into their constituent cells. Like Driesch before him, he observed that if cells were taken from early embryos, they grew into entire salamanders. Significantly, however, he also found that as time went on and embryos became more highly developed, single cells became less able to produce an entire embryo: they lost their totipotency and became more and more restricted in terms of the cell types they could produce.

There was, of course, still the nagging question as to whether somatic cells really did contain all the genetic material allotted to an animal in the fertilized egg or whether it had been diluted as suggested by Weismann. Here again, Spemann attacked the problem using some amazing technological wizardry. In 1923 he again employed a baby-hair lasso tied around a cell from an early salamander embryo. Gradually he tightened the noose around the cell until it became dumbbell shaped and the nucleus was forced into one side, leaving the other side empty. The egg then began to divide into cells, but this was restricted to the side containing the nucleus; nothing happened to the empty side. Spemann allowed development of this strange embryo to proceed until it had divided four times, yielding sixteen cells. He then relaxed his lasso, allowing a nucleus to slide back into the empty nondividing half-cell. Tightening the lasso once again, Spemann split the reconstituted cell off from the rest of the developing embryo. Amazingly, this single cell developed into an entire salamander embryo. This experiment clearly demonstrated that, at the sixteen-cell stage, the nucleus of an embryonic cell still contained all the requisite information for directing the development of an entire animal—at least this was what happened with salamanders. Of course, all the animals that had been made from individual embryo cells or fragments by Tremblay, Haeckel, Driesch, Spemann, and their many colleagues appeared in all respects identical to the animal from which the cells had been obtained; they were in fact "clones." We hear a lot about stem cells and cloning these days, but the roots of this topic stretch back well over a hundred years. Spemann won the Nobel Prize for his many contributions to the field of embryology.

Spemann's demonstration that the nucleus of a cell could be moved from one cell to another set the stage for a number of extremely innovative investigations starting in the middle of the twentieth century that greatly increased our understanding as to how gene expression is regulated during the process of development. The first of these studies was conducted by Robert Briggs and Thomas King in Philadelphia in 1952.[16] They began by taking a frog's egg, a frog oocyte; carefully puncturing it; and removing the nucleus. Now they took a cell from an early tadpole embryo, punctured it, and sucked its nucleus into a micropipette. They then injected this nucleus into the empty oocyte. Would the egg develop into a tadpole? It did. They did find, however, as Spemann had before them, that if they used older tadpoles as a source of nuclei, the egg would no longer develop properly, leading to the tentative conclusion that whatever changes had occurred within the nucleus during differentiation of the frog, they were not readily reversible after the early stages of development. It seemed that once cells were really differentiated, their nuclei could not be reprogrammed. This was to become a key issue.

The experiments carried out by Briggs and King and their predecessors had answered many questions and introduced important techniques, but some key issues remained. As we have seen, most of the successful experiments had used early embryos. However, there were indications that if a nucleus was transferred to an egg from a more differentiated tissue, then no viable clones were produced. This seemed reasonable. The cells from differentiated tissues—skin, liver, nerves, take your pick—are not normally observed to reverse their differentiated state. The latter stages of development do seem to be a one-way street. It appeared clear from the data that once a cell had differentiated, there was no going back again to the embryonic state. The genomes of these cells were in a state of lockdown. Following the earliest stages of embryonic development, cell differentiation seemed to be a completely unidirectional affair. One of the things this meant was that if you were ever going to clone an animal, you would have to use early embryonic cells as a source of their genetic material, as nuclei from differentiated tissues would not work.

However, not everybody was so sure about this. In 1956 John Gurdon, a new graduate student at Oxford University, decided to investigate things further.[17] He used the frog *Xenopus laevis* for his studies and began by successfully repeating the results that Briggs and King had obtained, transferring the nucleus of early embryonic cells to an enucleated egg, and obtaining viable tadpoles. Then he revisited the idea that nuclei from adult cells might also be reprogrammed. To do this, he took the nucleus from a mature frog's epithelial cell and transferred it to an enucleated oocyte. To everybody's surprise, and in contrast to the results reported by Briggs and King, in a number of cases he was able to generate perfectly good tadpoles. This was an enormous advance. In later years, he was able to achieve the same thing by transferring nuclei from cells obtained from other mature frog tissues as well. Gurdon had achieved a defining result in the history of biology: the idea that not only was the genetic material in all cells the same and possessed all the genetic information to produce an entire animal but also cells whose genes had been shut down during development could be reprogrammed back to the state of an early embryo. To be sure, the procedure that he used in the beginning, now named somatic cell nuclear transfer (SCNT), wasn't very efficient. Nevertheless, Gurdon was able to make multiple clones from a single frog. And of course, his experiments raised an obvious "what if" question. If it could be done with frogs, then why not with humans? This sounded like the type of science fiction that could be put alongside Mary Shelly's *Frankenstein* and H. G. Wells's *The Island of Dr. Moreau*. Gurdon's method was widely adopted and sparked investigations all over the world. The next step was to perform what is known as "reproductive cloning," that is, cloning complete animals, using SCNT from the cells of *adult* mammals. They wouldn't be humans, but it was a certainly a step along the path to human cloning.

It didn't happen right away. There were many stops and starts. In fact, it would take another forty years to clone mammals using SCNT from differentiated cells as Gurdon had achieved using frogs. It's not clear why. Many of the final steps were taken by scientists at the Roslin Institute for Animal Research at the University of Edinburgh in Scotland. Here, in 1996, Ian Wilmut, Keith Campbell, and their colleagues managed to transfer nuclei from embryonic sheep cells, which had been kept alive in cell culture dishes, into enucleated sheep eggs. The result was the birth of two lambs named Megan and Morag.[18] The ultimate experiment took place more or less by accident. The history of science is littered with stories of "happy accidents," but are they really accidents or just examples of capable investigators making the best of every situation? Be assured, it is the latter. In this case, the story is that the group of scientists at the Roslin Institute had prepared some nice sheep oocytes ready to receive nuclei by SCNT from cultured embryonic sheep cells. The cell preparation proved to be contaminated, but the scientists didn't want to waste their oocyte preparation. The only things around were some epithelial cells cultured from a six-year-old adult sheep. Should they try them instead? The scientists discussed the fact that this procedure couldn't possibly work with adult sheep cells based on everybody's previous experience, something that often happens in science. You all discuss why such and such an experiment couldn't possibly work, and then you do it anyway. It's just for fun, if you like—just to take a look and see what happens. Sometimes you get a surprise—your pessimism was misplaced; you just didn't really know enough about what was going on. And that's what happened in this case: 277 oocytes were injected with nuclei from the cultured adult sheep cells. Surprisingly, some of them did turn into embryos. Karen Walker, one of the scientists who performed the oocyte injections, related how she kept the developing embryos warm when taking them from one building to another on frosty Scottish mornings by putting them in tubes and stuffing them down her bra. Eventually, 29 of the resulting embryos were transplanted into surrogate ewes. Most of them were aborted, but there was one! The scientists watched the pregnant ewe with bated breath. After 120 days, Dolly the sheep was born, the first mammal to be completely cloned from a differentiated adult cell.

Dolly's story became famous all over the world. She is certainly one of the greatest animal superstars in history because she had proved a point. She showed what could be done. It was a "proof of concept" experiment, and it wasn't that long before the Scottish scientists did something even more extraordinary. By this time, it had become possible to start manipulating the DNA of an animal's genome using enzymes. The enzymes acted like scissors—you could snip a piece of DNA out here or add another piece there. The Scottish researchers wanted to demonstrate the potential utility of their work, so they took cultured sheep cells and, like a molecular Dr. Moreau, inserted a human gene into their DNA. The gene was for the protein Factor IX. This protein is a vital component of the family of proteins that clot blood. However, it's deficient in some humans producing a type of hemophilia. It is also not easy to purify this protein, and supplies are difficult to obtain. Now, the scientists took nuclei of edited sheep cells and inserted them into oocytes. Embryos developed, and ewes were born—Polly and Molly. Both sheep produced buckets of human Factor IX protein in their milk. Now, perhaps, mankind would have a limitless supply of this or other proteins that are therapeutically important.

Of course, since the birth of Dolly, Polly, and Molly (and let's not forget Holly and Olly), cloning animals—cows, monkeys, goats, and Snuppy the dog—using SCNT has become a relatively common procedure. There have even been attempts at preserving endangered or extinct species—the ancient gaur and mouflon. In 2009, the last remaining Pyrenean Ibex, or bucardo, was found dead. Anticipating this eventuality, scientists had previously obtained skin samples from the animal, which they kept frozen. They transferred the nucleus of one of these skin cells to the enucleated egg of a goat, a closely related species. An embryo started to develop. Implantation of the resulting embryo into a surrogate mother goat resulted in the birth of a kid. The kid eventually died because of a lung problem, but the experiment clearly indicates how close we are to success in such an enterprise. The dodo, it seems, may not be quite as dead as we had previously believed.[19] There have also been several successes including the recent cloning of the endangered black-footed ferret,[20] and cells from many other endangered species are being "banked" in preparation for future disasters. All these cloned animals seem fine, not some kind of Frankenstein creation as some had feared. Dolly developed arthritis and died young. Some said it was a sure sign that all was not well. But then four more animals were cloned from her cells—Debbie, Denise, Diana, and Daisy.[21] Only Debbie developed a little bit of arthritis. The four sheep all lived to ten years old and seemed perfectly normal. (However, they were euthanized when they reached the age of ten as this is rather old for sheep and the scientists wanted to avoid a situation where they developed significant age-related disorders.[22]) This illustrates the powerful potential of such an experiment. The four "Nottingham Dollys" were all identical clones, and so figuring out why one gets a disease like arthritis and not the others can provide us with an excellent opportunity for discovering how environmental factors like trauma can influence the development of pathology in otherwise identical individuals.

Since the birth of Dolly, our ability to use SCNT to clone different types of animals has increased greatly in terms of sophistication and efficiency. The original cloning of Dolly was a tour de force and represented a positive result obtained in the midst of innumerable failures. Since that time, animal cloning has become much more efficient and reproducible and is now considered to be a normal part of agricultural science, where it is applied to the production of livestock. A vast array of animals including deer, horses, rabbits, cats, rats, mice, dogs, and monkeys have also been successfully cloned. Whereas the production of Dolly was more or less a hit-or-miss affair, nowadays methods like these are routine in farming and agricultural circles.

It seems, then, that we have mastered many of the secrets of embryological development of animals, and, with the help of molecular genetic techniques, we can produce animals with particular properties essentially from scratch. But—here is the big question—can we do the same with humans? Can we produce a human Dolly? Clearly, we have learned a great deal about the growth and development of embryos from animals. How much of this information is applicable to humans? What information is available to us that would allow us to construct human embryos from scratch, and, if we can do this, what exactly could these be used for? Embryologists have had their own model organisms that they have studied over the years. Traditionally, as we have seen, these have been things like sea urchins, frogs, chickens, and, more recently, mice. How closely does the development of a human embryo resemble that of a mouse? The story here is reminiscent of the story of the cardiovascular system. There are many general similarities but also enough differences that our precise knowledge of the development of a human embryo can only be obtained by studying a human.

Cloning Humanity

Mammalian embryogenesis begins with the fertilization of an egg cell, an oocyte, by a sperm cell. As the oocyte and the sperm contain the genetic information from the mother and father, respectively (they are haploid or monoploid), the zygote will contain copies of both—it is "diploid." The zygote has the ability to produce every cell of the future organism, together with all of the important extraembryonic structures like the placenta and yolk sac. The zygote begins to divide, forming two cells, then four cells, and so on. This process can be achieved in a cell culture system and forms the basis for in vitro fertilization (IVF). Early development of the embryo seems similar in different mammalian species such as a human and a mouse. However, once the embryo begins to establish interactions with the maternal uterine endometrium (implantation), major structural changes take place, and mouse and human embryogenesis diverge significantly.

What happens if you remove one of the early cells from a human or mouse embryo and attempt to grow it by itself? The general observation is that, as observed with experiments going back to the nineteenth century, the cell can continue to divide and, under the appropriate culture conditions, go on to form the beginnings of an embryo. Interestingly, analysis of these early cells using the latest techniques has shown that they aren't completely identical. Even going back to the earliest cell divisions when there are just four cells, the biochemical nature of each cell is subtly different.[23]

After about five days, the human embryo forms a structure that illustrates the fact that different cells are beginning to make their own way in the world. This structure is a hollow ball of cells called a blastocyst. The outside of this structure will form the placenta, and the group of cells on the inside—the "inner cell mass" (ICM)—will form the entire resulting human. Interestingly, individual cells plucked from the ICM can act as stem cells that have the ability to form all the different human tissues. Hence, they are known as human embryonic stem cells (hESCs) and permanent cell lines can be established from them. These cell lines can be grown indefinitely in their undifferentiated state and, when provided with the appropriate biochemical instructions, be transformed into mature tissues.

The next extremely dramatic stage of embryonic development involves the firm attachment of the embryo to the uterus, at which point a process called gastrulation occurs when the embryo forms three layers—ectoderm, mesoderm, and endoderm. In nearly all animals, from flatworms to primates, these cell types have been found to have a similar role: cells of the ectoderm become skin and nervous system; cells of the mesoderm form skeletal muscle, bone, connective tissue, the heart, and the urogenital system; and cells of the endoderm form the lining of the digestive tract, as well as that of the lungs and thyroid. At this point, the embryo has become a "fetus."

Over the last few years, some groups of researchers have managed to coax human embryos in cell culture to develop from zygotes to the point of implantation to some synthetic surface, when cells begin to differentiate into the appropriate three layers that normally occur during gastrulation.[24] However, the experiments were all terminated after thirteen days. The reason for this is the "fourteen-day" rule, which was agreed upon internationally.[25] It was established in 1979 in the United States and in 1984 in the United Kingdom in an attempt to satisfy the different stakeholders in the debate about the potential use of human embryos for research purposes following the original development of procedures for IVF in 1978. The limit of fourteen days originated because of key events in the formation of the gastrula around the fifteenth day of human embryo development. It was felt that these changes were sufficient to designate the fetus as an "individual." This restriction means that many interesting and medically important events in embryogenesis, including the study of gastrulation, establishment of the neural plate and tube, major organs, and body axes, and the origin of primordial germ cells (PGCs), cannot be studied in culture systems of this type. Organogenesis in humans begins around twenty-one days, which is a period that is often critical for the teratogenic actions of different drugs and chemicals and would therefore be of great interest to study. Recently, the successful culture of cynomolgus monkey embryos for up to twenty days beyond the gastrula stage was reported. Other recent studies have reported the successful culturing of mouse embryos from day five up to day eleven in a newly devised culture chamber without a uterus, at which point they are well along the road to developing identifiable limbs and organs (the entire gestational period for a mouse is only twenty days[26]). Results such as these suggest that long-term culture of human embryos may also be feasible if it was permitted. Indeed, the entire matter of the fourteen-day rule is currently under review,[27] and very recently an oversight committee has ruled that this period should be extended.[28] This, together with ever-improving techniques for culturing embryos,[29] will certainly lead us into the *terra incognita* of human embryological development.

From the time Dolly was cloned, there has been great interest in the possibility of cloning a human using similar methods, and sure enough, claims of success in doing this followed soon after Dolly's birth. In 1998, the appropriately named Richard Seed, a physicist, generated a great deal of publicity by claiming to have initiated the cloning of humans, in particular of himself and his wife, Gloria, but nothing ultimately transpired. Success in cloning human embryos was first reported by a South Korean scientist in 2004. It caused a sensation, yet in the end, it turned out he had cheated and hadn't done what he said he had. He was dismissed in disgrace. Around the same time two Italian obstetricians, Severino Antinori and Panayiotis Zavos, announced that they had successfully started multiple pregnancies from cloned humans, but here again nothing ultimately happened. And then there were the intriguing claims of a religious cult called the Raëlians that declared that an alien intelligence had commanded them to clone humans. In 1997, about three months after Dolly's birth was announced, the group created a biotech company called Clonaid aimed at human cloning, a project they pursued until they were ordered to stop by the Food and Drug Administration (FDA). Undaunted, Clonaid relocated to the Bahamas. On December 27, 2002, the group announced that the first cloned baby, a girl named Eve, had been born. By 2004, Clonaid claimed to have successfully engineered the birth of fourteen human clones. Regrettably the Raëlians have never allowed anybody to have access to these individuals, so we only have their word for it. Generally speaking, the Raëlians' claims have not been given any credence whatsoever—but I suppose we don't really know, do we? Eventually, in 2013, genuine human embryos were cloned by several groups and hESC lines established from them.[30] This took a while. It seemed that some obscure factor was needed in the culture medium to get the entire thing to work properly. Eventually it was discovered to be caffeine! Makes you think, doesn't it? What you do is this (it's just like making the human equivalent of Dolly): Take a skin cell from a particular individual, transfer the nucleus of this cell into a donated human egg that has been enucleated, and then stimulate this egg using certain chemicals or electric shocks, which will trigger it to begin developing into an embryo. When it reaches the blastocyst stage, one can remove cells from the ICM and put them into cell culture. These are hESCs that will potentially be able to generate any of the tissues of the person who donated the fibroblast. If the individual carries disease-related gene mutations, then you will have created a line of hESCs that possesses this trait, something that might be useful when studying the disease. However, all of this is anything but straightforward. Making stem cells from a blastocyst results in death of the embryo and its life-giving potential. If the blastocyst was implanted into a surrogate mother, it might be carried to term and the resulting baby would be a clone of the fibroblast donor. Naturally, such possibilities have been hotly debated. The use of hESCs has always been associated with an intense debate about ethics, and this, rather than technical issues, has often stood in the way of their general use.

Achieving the goal of human reproductive cloning seems like one of the pinnacles of all stem cell biology, or any kind of biology for that matter. The idea that we can create a human from scratch is somehow very exciting and fulfilling from the scientific point of view. Aren't the secrets of this process the basis of all human existence? If this is something we can do at will, then surely we have become gods! We can both create our own species and, by recreating ourselves in some form, perhaps cheat death. The uses and abuses of these methods have been widely discussed, as they should be. The Nobel Prize–winning author Kazuo Ishiguro in his novel *Never Let Me Go* discussed one possibility: clone humans and then use their tissues for transplant surgery or research purposes. In the book, this has become the accepted way of doing things. People don't even give it a second thought, and of course, one can imagine a "slippery slope" where things start out one way and end up being something we shudder to think about.

Nevertheless, there are many more mundane reasons for using human tissues if they were available that don't involve the vivisection of human beings or destroying embryos. So, let's take a step back for a moment. The fact is that as cells differentiate and organs develop, they don't all lose their original stem cell identity. Most tissues, even in adults, contain populations of cells that have a more restricted regenerative potential. These cells would not be able to generate an entire animal but would be able to generate the cells of the tissue from which they were sourced. As opposed to pluripotent stem cells (PSCs) like ESCs, these are called adult stem cells (ASCs) and represent stem cell populations with restricted potential that could be used for producing particular tissues. Things like this have been done; consider meat.[31] For several years now companies have been working on the idea that they could prepare meat that didn't involve killing animals. A lot of people, including me, don't want to eat meat for that very reason. There is nothing wrong with meat per se as food, but the process of obtaining it through the terrible mistreatment of animals makes it unacceptable for numerous reasons. If no animals were involved, I would imagine that for many people eating meat would not be a problem. If one takes ASCs from bovine muscle, for example, one can coax them into making beef muscle using the appropriate cell culture conditions. We don't need to make an entire animal. Indeed, we don't need to make any kind of animal. Products like this are apparently going to show up in our grocery stores very soon, even though there will be other issues and questions, no doubt, like whether this new meat is kosher/halal?

When it comes to stem cells, ESCs and ASCs are only the beginning of the story. Around 2000, several groups of scientists began to investigate ESCs to see if they could identify the factors that were essential for keeping them in their pluripotent state and prevent them from differentiating into tissues. Gradually, important factors began to emerge. One of the first of these was a transcription factor (a gene that controls gene transcription) called Nanog (from the Gaelic Tir Na Nog—land of youth). Then others followed. The laboratory of Shinya Yamanaka in Japan began a systematic investigation of the transforming power of groups of these transcription factors.[32] First using fibroblasts from embryonic mice, they expressed up to twenty-four different transcription factors at the same time—a considerable technical feat—and it worked! The mouse fibroblasts were transformed into cells that seemed absolutely like mouse ESCs. It was like molecular alchemy. Then Yamanaka and his colleagues began to try winnowing away the list of twenty-four factors and arrived at an essential group of four proteins that could "transdifferentiate" fibroblasts into bona fide stem cells with developmental potential equal to that of an ESC. Would it work using human cells? It did; a group of four proteins did the trick. Yamanaka and his colleagues named these cells "induced pluripotent stem cells" (iPSCs). The world quickly took notice. Yamanaka and Gurdon shared the Nobel Prize for their work on cell reprogramming. Obviously, there are some things about iPSCs that are particularly attractive. They are relatively easy to prepare. You can start with a fibroblast or indeed any other easily obtained cell and generate patient-specific stem cells that can then be transformed into any type of tissue. Basically, iPSCs and ESCs are equivalent in most respects, but to make ESCs you need an enucleated oocyte, and the technique is generally more challenging. Nevertheless, now we have two methods for making human PSCs. Naturally, there is presently a huge effort aimed at trying to figure out which types of stem cells are "better" and for what purposes. However, the fact that we can now all generate our own stem cells in a readily reproducible manner has an enormous number of potential implications for science and society. So, as we can see, the differentiation of somatic cells in tissues wasn't irreversible after all. Like nearly everything else, it just boiled down to a set of biochemical reactions, and if you knew which ones were important, with a bit of tinkering you could reverse biological time.

The availability of ESCs and iPSCs has suggested a general strategy for biological experimentation. hESCs and human induced pluripotent stem cells (hiPSCs) can generate all the different tissues that make up a human being. Hence, once these cells are growing in a culture dish, all you need to do is find out how to treat them, and they will develop into the types of cells you are interested in. These could be heart cells, nerve cells, bone cells, or virtually any other kind of cells. In fact, even mouse egg (germ) cells have been made this way and turned into baby mice.[33] Indeed, by treating these cells in the correct way using the appropriate growth factors and synthetic culture conditions, they can be differentiated into structures that resemble blastulas (blastoids) or gastrulas (gastruloids), and very recently structures resembling complete mouse embryos have been produced from ESCs if they are cultured with other types of stem cells that provide support tissues. No egg, no sperm-no problem! It is considered likely that making these "embryoid" structures from human ESCs will be achieved in the near future.[34] These methods will potentially enable scientists to investigate human embryogenesis beyond the fourteen-day barrier.[35]

It seems certain that step by step, starting with an hESC or a hiPSC, we are developing the technology for building a completely viable human embryo, which could one day result, ethical considerations aside, in a human being. It really does resemble Huxley's *Brave New World*, in which embryos were developed ex utero in special incubators, and there is no reason we would not be able to achieve such a thing in the future. Indeed, recent studies have also succeeded in making human endometrial tissues from stem cells, and it is highly likely that in the not-too-distant future we will be able to use stem cells to generate human embryos and human wombs that will carry them to term without the participation of a traditional mother and father.[36] There is no magic in this, just science. And perhaps we will not stop with human beings. Some studies have demonstrated that blastocyst-like structures can be made of cells from more than one species: pig or cow embryos containing human cells, rat and mouse hybrid embryos, and, most recently, chimeric monkey/human structures.[37] Where this line of experimentation will finally lead is not yet clear.

The Organoid Revolution

What if we are not developmental biologists and our ultimate aims are not understanding embryogenesis? As discussed above, the types of tissues we might potentially make from stem cells depends on understanding what "factors" to put into the culture medium. These may be things like specific growth factors and cytokines or special surfaces that help tissues to grow properly. And, of course, there are a huge number of human pathologies that need new tissues for repair purposes. Consider a neurodegenerative disease where we need to replace nerve cells, and not just any old nerve cells, but very specific populations. These might be the dopamine-utilizing cells from the substantia nigra in Parkinson's disease or motoneurons in amyotrophic lateral sclerosis (ALS). Such a thing won't be easy. Not only must the correct cells be grown in culture, but also they must then be persuaded to regrow within the patient to produce the correct synaptic connections. Nevertheless, results such as these are certainly no longer considered unattainable. Or consider type 1 diabetes, in which the insulin-secreting cells of the pancreas have been destroyed, or macular degeneration, where cells in the retina have been lost. We should be able to make more of the appropriate cells and replace them, and there are already numerous reports showing that therapies like these may be possible. Procedures like these are known as "therapeutic cloning"; the goal is not to produce an entire human being, just defined cell types for repair purposes. Also, note that if the stem cells used to generate different tissues are prepared from the particular patient involved, they will have many of the properties that define that individual genetically, so the immune system will be less likely to reject them—an important consideration. Indeed, stem cell–related methods have the potential for producing a biology that reflects the whole of human diversity, a huge advance on what can be achieved by traditional animal experimentation.

We might also consider another potential use for these technologies, one that goes beyond preparing a single type of cell. Why not create entire human organs from scratch for performing experiments and developing drugs? Here, we might imagine an approach that is quite sophisticated, one that doesn't involve destroying a human being. For example, if we want to work on joint physiology because we are interested in the pathogenesis of rheumatoid arthritis or osteoarthritis, we would need to use a joint preparation that consisted of bone, muscle, cartilage, synovium, immune cells, blood vessels, and innervating neurons. This might seem like a difficult thing to construct from scratch, but it turns out that doing things like this these days is quite feasible. We have already discussed the fact that stem cells like hESCs and hiPSCs can create multicellular structures such as blastulas or gastrulas, which are then termed "blastoids" or "gastruloids."[38] When similar protocols are employed using stem cells to construct other complex organs and tissues, these structures are called "organoids."[39] Starting with the appropriate stem cells, one can produce an entire organ system. Rather than simple layers of cells grown in two dimensions in a culture dish, organoids are three-dimensional self-organizing culture systems that are extremely similar to genuine human organs. Human organoids are usually generated from hiPSCs or hASCs, sourced from the relevant human tissue and grown under conditions that mimic normal human development. Organoids contain the multiple cell types that are found in the complete organ.

Nowadays it is relatively easy to see how well an organoid resembles a particular tissue by employing parallel advances in nucleic acid sequencing and bioinformatics. These techniques allow you to obtain the sequences of all the different mRNA molecules in a single cell, a procedure known as single-cell RNA sequencing, or scRNAsq. This kind of information helps precisely identify a cell—rather like a bar code identifying a particular product in a store. If you take all the cells from an organ like the heart and perform this experiment, you can then see exactly what kinds of cells are present based on the patterns of the mRNAs synthesized by each cell; you can then prepare a map displaying all the different cell types. These maps look very much like early paintings by abstract artists like Wassily Kandinsky, with different-colored blobs representing the different types of cells (Figure 6.4). Now you take an organoid that you have grown in cell culture that is meant to resemble the same tissue and perform exactly the same procedure, which will also result in a cell map. You can easily see how well the two cell maps correspond.

Many publications have now reported that cultured organoids grown from stem cells are often remarkably similar, if not identical, to authentic human tissues. When making an organoid, one can observe the course of normal development unfold before one's eyes. Human stem cells are sequentially exposed to a course of differentiation cues in order to simulate the stages of human development. During this process, differentiated hiPSCs or ASCs first aggregate to form an organ bud, followed by a mature organoid, which faithfully mimics the mature organ structure, including all the cell types that normally constitute the organ in question and the appropriate intercellular interactions between them. Another interesting aspect of this line of research is the preparation of special surfaces that allow stem cells to grow into

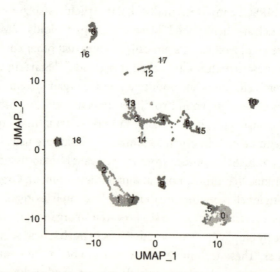

Figure 6.4 Separate cell clusters obtained by analysis using single-cell RNA sequencing from a mouse dorsal root ganglion. Each number represents a different cell type found in the tissue.

Image used with permission from R. Miller.

organoids that precisely resemble the shapes of normal organs. The application of all these new technologies has resulted in organoids that can exactly represent human organs grown completely from scratch.

It is hard to describe how quickly this field of science is moving forward. Nearly every issue of a top scientific journals these days contains one or more striking publications describing a new type of human organoid, its properties, and how it can be used to help solve human-related problems. Success has been achieved in producing many types of organoids including skin, intestine, kidney, retinal, pancreatic, prostate, and liver organoids (Figure 6.5).[40] Human tear gland organoids have been produced and induced to cry in cell culture![41] Muscle strips can be removed from stomach organoids and made to contract in an organ bath,[42] the latest version of the isolated tissue preparations that were first generated by Albrecht Haller back

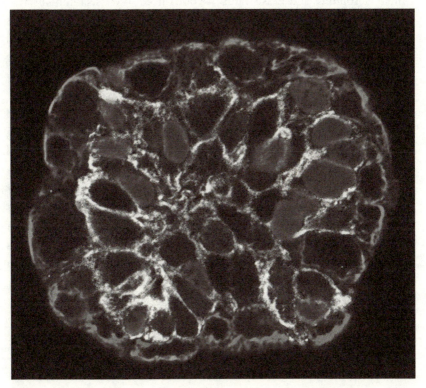

Figure 6.5 Human mammary gland organoid grown from stem cells, showing differentiated cell structure and organ morphology.
Image from Alamy.

in the eighteenth century (see Chapter 3). As far as the cardiovascular system is concerned, beating hearts can now be generated from people who have genetic deficits, producing an experimental model for finding out what exactly is going wrong and why. If we also consider tumors as organs, which in many respects they are,[43] they can also act as a source for unique types of organoids and used to analyze the factors that control their abnormal growth. Even tissues as complicated as the brain appear to form complex organoid-like structures. Brain organoids can be produced that resemble the entire brain or just a part of it such as the cerebral cortex.[44] And perhaps we shouldn't be surprised by all of this. All organs normally develop in the body according to certain rules and under the influence of sequential molecular cues. If we learn what those rules and influences are, then why shouldn't we be able to mimic them exactly and produce individual human tissues under conditions that suit us, rather than having to source them from intact human beings? Unlike Herophilus and Erasistratus, we don't need to vivisect humans anymore to examine human biology. Just like any other biological process, this is another example of learning the appropriate rules of biochemistry and cell biology and putting them into action. A perfect organoid represents nothing more or less than an entire human organ generated from stem cells that can be maintained in a culture system. If we consider organoids as substitutes for genuine human organs, then might they not be used for the testing and development of therapeutics that are directed to humans rather than using animals? Such a thing would be revolutionary, and the revolution is well underway.

Organoids are already being used for drug development and other procedures that are not just specific for humans but for individual humans at that. For example, gut organoids have proved to be powerful tools in personalized medicine, allowing scientists to study the effectiveness of drugs in people with cystic fibrosis (CF), who have genetic defects that affect ion channels and disrupt the movement of water in and out of the cells lining the lungs and intestine.[45] Researchers can take rectal biopsies from patients with CF, use the cells to create personalized gut organoids, and then apply a potential drug to see if it will reverse the symptoms of the disease in that individual. This is important, as drug responsiveness is not the same in different patients with CF; it very much depends on the particular mutation they carry. Here, then, is an example of the personalized medicine of the future. No animals required!

Furthermore, why shouldn't organoids replace animals and be used by the pharmaceutical industry for drug safety testing? This is another revolution that is currently underway. Using human organoids for ADMET (absorption, distribution, metabolism, excretion, toxicology) studies will help to predict metabolic mechanisms and ascertain key safety and efficacy measures before commencing clinical trials. Kidney organoids have been produced using hiPSCs.[46] They contain all the cells normally found in a human kidney organized with the correct tissue morphology. Because the kidney plays a key role in drug metabolism and excretion, a kidney organoid is useful for testing drug candidates for toxicity before they reach clinical trials.[47] Kidney and intestine organoid models can be of great value when investigating the specifically human metabolism, transport, and toxicity of drugs. Researchers speculate that other human organoids, such as heart and liver, could similarly be used to screen drug candidates for toxic effects, resulting in human-specific data that is not possible to obtain with animal testing.

Findings such as these highlight the future potential of organoids for drug development by the pharmaceutical industry. Because these systems closely resemble real human tissues, they could help predict drug responses early in development and offer numerous possibilities for modeling many diseases in the future. Use of such techniques should yield information that is much more relevant to humans than the use of animal-based tests, which, as we have discussed are usually extremely unreliable. Indeed, the first new drug application to be developed this way has just been reported. Researchers in Israel used kidney organoids on specially designed microchips endowed with micro-recording devices to investigate how the commonly used anticancer drug cisplatin can produce kidney toxicity through increased buildup of fat.[48] They were able to reproduce this problem using their microchip technology by treating kidney organoids with cisplatin, and furthermore, they were able to show that a drug called empagliflozin could reduce the fat buildup. Finally, they were also able to elucidate the drug's mechanism of action. This data has now been presented to the FDA so that the drug can be approved for this purpose. Absolutely no animals were used in these experiments, only human-derived tissues. It is therefore abundantly clear that the future of drug development is already with us, and the big pharmaceutical companies will be able to perform all their requisite safety testing using cutting-edge human-specific technology and not woefully out-of-date, unreliable, and extremely crude animal testing. Better, cheaper, more predictive, more reliable methods are already with us. Everybody wins. Dumping the use of animal testing by industry will remove millions of animals from death row.

Furthermore, it isn't only scientists in industry who will benefit from using organoids instead of animals. Similar considerations apply to researchers in universities carrying out "basic" research. If the goal of their research is translation to humans, then working on human tissues will be much more relevant. Institutions all over the world have begun to organize banks of human stem cell lines from individuals who carry many interesting genetic traits, allowing investigators to really work on projects that reflect human diversity, which is something that animals simply cannot do.

Organoids have already been used in some important scientific investigations, such as understanding the pathological actions of the Zika virus (ZIKV).[49] ZIKV infection of humans was first identified in 1952. The infection is spread by mosquitoes, and the resulting illness in adults is typically mild with fever-like symptoms that last for a few days. However, in 2016, ZIKV was declared a public health emergency by the World Health Organization based on observations that there was an association between babies born with small brains (microcephaly) and ZIKV infection. Because there was no suitable model for studying ZIKV infection, it was modeled in stem cell–derived brain organoids designed to resemble cerebral cortical structures. These experiments were a great success. ZIKV infection was found to cause a reduction in the proliferation of infected neural progenitor cells and an increase in their death, something that was responsible for the development of abnormally small brains, one of the most obvious symptoms of ZIKV infection in children. It was also observed that the infected organoids upregulated expression of the innate immune receptor TLR3 (toll-like receptor 3), which was responsible for producing the observed dysregulation of the birth and death of nerve cells. This suggested that a drug that blocks TLR3 might be therapeutically helpful in treating ZIKV infection. Organoids have also been used to model the effects of other viral and bacterial pathogens including norovirus and clinical strains of rotavirus, two viruses that were difficult to cultivate and study until organoids became available as a tool. Recently, the use of organoids for investigating host/pathogen interactions has become of great importance in understanding the biology of the SARS-CoV-2 virus and how it produces disease during the coronavirus pandemic.[50] As exemplified by the ZIKV studies, a human model system is preferable to animal models when studying infectious disease pathogenesis because pathogens often have a narrow species or tissue specificity (tropism).

It is certainly more informative to study a human-specific virus using human tissues.

One potential drawback of organoid systems is the lack of the type of interorgan communication one sees in a live animal. Organs are normally connected by things like circulating blood and lymph, which are essential for maintaining overall coordination and homeostasis, reflecting Claude Bernard's *milieu interieur*. Fortunately, such problems can also be overcome. There are currently efforts to combine organoid research with advances in microfluidics and "organ on a chip" systems, allowing the generation of spatially separated organoids that can be connected to one another. The use of three-dimensional printing is another innovation, allowing the creation of three-dimensional scaffolds that can be seeded with organoid systems that are connected through microperfusion devices. Multiple organoids (assembloids) have been connected using such systems to study communication between the liver, pancreas, and gastrointestinal tract. In 2019, one group of scientists grew human brain organoids next to a mouse spinal column and back muscle. Nerves from the human organoid grew into the spinal column, and the muscles began to spontaneously contract.[51] More recently brain assembloids have been prepared by growing several parts of the brain together.[52] They organized themselves into a complex brain organoid that exhibited the expected emergent physiological properties. When the authors made the same organoids using stem cells from patients with Rett syndrome, the organoids displayed the epileptic seizures that are typical of the disease. Finally, the authors found that they could reverse these pathological changes with a drug, suggesting a therapeutic intervention for treating the disease.[53] Imagine all of this using human tissues from patients who really have diseases of the nervous system—not animal models that are inadequate substitutes. It is now easy for us to imagine sophisticated setups of interconnected human organoids that represent the construction of any portion of a complete human being, the constituent parts all throbbing in unison—something that once might have been thought of as science fiction, but not anymore.[54]

Here, then, we would have created the biological essence of a human generated entirely from stem cells. Indeed, because of the nature of organoids and their increasing complexity, it is possible that in the future sophisticated brain organoids might even develop some form of "intelligence," something that bioethicists have already started to think about. If they do, how should the law view them? Will they qualify as "persons" and be covered by laws protecting "personhood"? All of this may seem premature, but in some experiments, investigators have been able to record brain activity following the stimulation of the eye cup regions of brain organoids with light.[55]

Indeed, organoids can contribute to the most fundamental basic scientific inquiries, such as answering the question "What does it take to be a human?," as opposed to something else such as a chimpanzee. A recent study in the journal *Nature* compared the development of cerebral cortex organoids from humans and chimps and presented data detailing gene expression patterns that occurred during this process. The authors were able to demonstrate interesting differences in the temporal flow of gene expression that characterized uniquely human aspects of human cortical development.[56] And it's not only the differences between species but also the differences between sexes that are amenable to this approach; brain organoids have proven to be excellent models for investigating the basis for the sexual dimorphism of the human brain.[57]

Even more remarkable, perhaps, when considering the biology of humankind are studies that compare modern humans with Neanderthals. Although Neanderthals only exist in our past, they are just half a step away from being "us," closer indeed than a chimpanzee. And although there are no Neanderthals around these days, there is Neanderthal DNA that has been obtained from ancient remains. The sequences of some Neanderthal genes show small differences from the genes of *Homo sapiens*. An experiment can therefore be performed that takes advantage of the entire arsenal of modern nonanimal biology, using CRISPR gene editing to replace the Neanderthal sequence of a gene, in this case the gene NOVA1, in a brain organoid grown from the cells of a modern human. How does this brain now differ? The experiments are ongoing, but the brain organoids certainly *look* different. It is by performing studies such as these that we may eventually come to understand the final steps evolution took when producing modern humans.[58]

It seems, therefore, that there are an ever-increasing number of options available to researchers that they can use in their experiments instead of using animals. Not only that, but these methods have many advantages over animal systems, given the fact that they are derived from humans and are therefore directly relevant to the study of human beings. The use of increasingly sophisticated gene sequencing and stem cell technologies represents a scientific nexus that is telling us a great deal about exactly what makes up a human as opposed to a monkey or a mouse, information that can make all the difference when we are designing new therapies for human diseases by allowing us to generate specifically human tissues for experimental purposes. In a rapidly increasing number of instances, animal models are simply no longer necessary, and, to repeat, organoids and related systems can be made to reflect all of human diversity, which is such an important issue in medicine and therapeutics. And of course, it isn't just stem cell–related methodologies that are important these days. Technologies such as live imaging of human subjects can reveal the activity of human tissues such as the heart and brain in real time (see Chapter 5). In addition, advances in the study of human psychology and behavior, robotics, and artificial intelligence make such investigations much more relevant to humans than investigating monkeys or mice.

The most embarrassing thing for many scientists is that work on animals is starting to look desperately old-fashioned. The cutting edge has moved on, and animal experimentation is now behind the curve—and believe me, the worst and most humiliating thing a scientist wants to acknowledge is that their work isn't up to date. Science lives and dies by being ahead of the curve, by moving the boundaries forward. If a scientist isn't up to date, they are functionally dead. Of course, scientists who are well established in their careers and have made extensive use of animals in the past feel threatened by this kind of talk and usually defend themselves by saying that some aspects of human biology will always require animal models and will never be possible using aspects of stem cell/organoid or other technologies. But clever young scientists at the start of their careers don't feel that way and are moving the field forward so quickly that the specter of animal experimentation is rapidly disappearing in the rearview mirror.

The question, therefore, is why are scientists still conducting experiments on animals, and how quickly will the situation change? The answer to this question is complicated, and we will discuss it further. But honestly, there isn't any real reason to do research using animals these days, and perhaps we have known this for a very long time.

Experiments should be carried out on the human body. If the experiment is carried out on the bodies of other animals it is possible that it might fail for two reasons: the medicine might be hot compared to the human body and be cold compared to the lion's body or the horse's body.

The second reason is that the quality of the medicine might mean that it would affect the human body differently from the animal body.

Avicenna's *Canon of Medicine*, tenth century CE

Notes

1. Zvonimir Vrselja et al., "Restoration of Brain Circulation and Cellular Functions Hours Post-Mortem," *Nature* 568, no. 7752 (April 2019): 336–343, doi:10.1038/s41586-019-1099-1.

2. George K. Michalopoulos, "Liver Regeneration after Partial Hepatectomy," *American Journal of Pathology* 176, no. 1 (January 2010): 2–13, doi:10.2353/ajpath.2010.090675.

3. Aristotle, A. L. Peck, and Aristoteles, *History of Animals. 1: Books I—III, with an English transl. by A. L. Peck*, repr., Aristotle 9 (Cambridge, MA: Harvard University Press, 2007), bk. 2, chap. 17.

4. Hans R. Bode, "The Interstitial Cell Lineage of Hydra: A Stem Cell System That Arose Early in Evolution," *Journal of Cell Science* 109, no. 6 (June 1996): 1155–1164, doi:10.1242/jcs.109.6.1155; Howard M. Lenhoff and Sylvia G. Lenhoff, "Challenge to the Specialist: Abraham Trembley's Approach to Research on the Organism—1744 and Today," *American Zoologist* 29, no. 3 (August 1989): 1105–1117, doi:10.1093/icb/29.3.1105.

5. Richard J. Miller, This section is adapted from *I Am a Mutant: A Genetic Odyssey* (Chicago: CreateSpace Independent Publishing Platform, 2017), 237–263.

6. Bode, "The Interstitial Cell Lineage of Hydra."

7. C. E. Dinsmore, "Urodele Limb and Tail Regeneration in Early Biological Thought: An Essay on Scientific Controversy and Social Change," *International Journal of Developmental Biology* 40, no. 4 (August 1996): 621–627; Elizabeth R. Zielins et al., "The Role of Stem Cells in Limb Regeneration," *Organogenesis* 12, no. 1 (January 2016): 16–27, doi:10.1080/15476278.2016.1163463.

8. Laura Otis, *Müller's Lab* (Oxford; New York: Oxford University Press, 2007).

9. Klaus Dose, "Ernst Haeckel's Concept of an Evolutionary Origin of Life," *Biosystems* 13, no. 4 (January 1981): 253–258, doi:10.1016/0303-2647(81)90005-8; Michael K. Richardson and Gerhard Keuck, "Haeckel's ABC of Evolution and Development," *Biological Reviews of the Cambridge Philosophical Society* 77, no. 4 (November 2002): 495–528, doi:10.1017/S1464793102005948.

10. Ariane Dröscher, "Images of Cell Trees, Cell Lines, and Cell Fates: The Legacy of Ernst Haeckel and August Weismann in Stem Cell Research," *History and Philosophy of the Life Sciences* 36, no. 2 (October 2014): 157–186, doi:10.1007/s40656-014-0028-8.

11. T. W. Holstein, E. Hobmayer, and U. Technau, "Cnidarians: An Evolutionarily Conserved Model System for Regeneration?," *Developmental Dynamics* 226, no. 2 (February 2003): 257–267, doi:10.1002/dvdy.10227.

12. Ernst Haeckel, *Art Forms in Nature: The Prints of Ernst Haeckel* (Munich; New York: Prestel, 1998).

13. Yawen Zou, "The Germ-Plasma: A Theory of Heredity (1893), by August Weismann," *Embryo Project Encyclopedia*, January 2015, https://embryo.asu.edu/pages/germ-plasm-theory-heredity-1893-august-weismann.

14. "Developmental Biology: Observing Evolution in Fast Forward," Institution for Science Advancement, February 2018, http://ifsa.my/articles/developmental-biology-observing-evolution-in-fast-forward.

15. Andreas-Holger Maehle, "Ambiguous Cells: The Emergence of the Stem Cell Concept in the Nineteenth and Twentieth Centuries," *Notes and Records of the Royal Society* 65, no. 4 (December 2011): 359–378, doi:10.1098/rsnr.2011.0023.

16. Ian Wilmut, Yu Bai, and Jane Taylor, "Somatic Cell Nuclear Transfer: Origins, the Present Position and Future Opportunities," *Philosophical Transactions of the Royal Society B: Biological Sciences* 370, no. 1680 (October 2015): 20140366, doi:10.1098/rstb.2014.0366.

17. Ruth Williams, "Sir John Gurdon: Godfather of Cloning," *Journal of Cell Biology* 181, no. 2 (April 2008): 178–179, doi:10.1083/jcb.1812pi.

18. Wilmut, Bai, and Taylor, "Somatic Cell Nuclear Transfer"; "Dolly, Polly, Molly, Megan and Morag," *Towards Dolly* (blog), July 2013, https://libraryblogs.is.ed.ac.uk/towardsdolly/2013/07/05/dolly-polly-molly-megan-and-morag/.

19. Julianna Photopoulos, "Jeanne Loring: Reversing Extinction," *Nature* 597 (September 2021): S18–S19.

20. "News at a Glance," *Science* 371, no. 6532 (February 2021): 868–869, doi:10.1126/science.371.6532.868.

21. Cecile Borkhataria, "On the 20th Anniversary of Dolly's Birth, Researchers Reveal They Are Putting the Four Other Animals Cloned from the Same Cell Line to Sleep as There Is 'No Scientific Merit' in Keeping the Nine Year Old Sheep Alive," *Dailymail.com*, February 2017, https://www.dailymail.co.uk/sciencetech/article-4250664/Dolly-cloned-sheep-s-four-clones-set-euthanized.html; Sharon Begley, "Dolly the Sheep Died Young, but Her Cloned Sisters Are Still Alive and Kicking," *Statnews*, July 2016, https://www.statnews.com/2016/07/26/dolly-sheep-clone-aging/.

22. Begley, "Dolly the Sheep Died Young, but Her Cloned Sisters Are Still Alive and Kicking"; Borkhataria, "On the 20th Anniversary of Dolly's Birth, Researchers Reveal They Are Putting the Four Other Animals Cloned from the Same Cell Line to Sleep as There Is 'No Scientific Merit' in Keeping the Nine Year Old Sheep Alive."

23. Mubeen Goolam et al., "Heterogeneity in Oct4 and Sox2 Targets Biases Cell Fate in 4-Cell Mouse Embryos," *Cell* 165, no. 1 (March 2016): 61–74, doi:10.1016/j.cell.2016.01.047; Matteo A. Molè, Antonia Weberling, and Magdalena Zernicka-Goetz, "Comparative Analysis of Human and Mouse Development: From Zygote to Pre-Gastrulation," *Current Topics in Developmental Biology* 136 (2020), 113–138, doi:10.1016/bs.ctdb.2019.10.002.

24. Embryo assembly 101. Researchers are starting to demystify the earliest stages of human development, edging right up to an ethical red line.
   Helen Shen, "The Labs Growing Human Embryos for Longer Than Ever Before," *Nature* 559, no. 7712 (July 2018): 19–22, doi:10.1038/d41586-018-05586-z; Kate Williams and Martin H. Johnson, "Adapting the 14-Day Rule for Embryo Research to Encompass Evolving Technologies," *Reproductive Biomedicine & Society Online* 10 (June 2020): 1–9, doi:10.1016/j.rbms.2019.12.002.

25. Williams and Johnson, "Adapting the 14-Day Rule for Embryo Research to Encompass Evolving Technologies."

26. Alejandro Aguilera-Castrejon et al., "Ex Utero Mouse Embryogenesis from Pre-Gastrulation to Late Organogenesis," *Nature* 593, no. 7857 (May 2021): 119–124, doi:10.1038/s41586-021-03416-3.

27. Elizabeth Svoboda, "Frontier of Development," *Nature* 597 (September 2021): S15–S17, https://media.nature.com/original/magazine-assets/d41586-021-02625-0/d41586-021-02625-0.pdf.

28. Insoo Hyun et al., "Human Embryo Research beyond the Primitive Streak," *Science* 371, no. 6533 (March 2021): 998–1000, doi:10.1126/science.abf3751; Kelly Servick, "Door Opened to More Permissive Research on Human Embryos," *Science* 372, no. 6545 (May 2021): 894–894, doi:10.1126/science.372.6545.894.

29. Kendall Powell, "What's Next for Lab-Grown Human Embryos?," *Nature* 597, no. 7874 (September 2021): 22–24, doi:10.1038/d41586-021-02343-7.

30. David Cyranoski, "Human Stem Cells Created by Cloning," *Nature* 497, no. 7449 (May 2013): 295–296, doi:10.1038/497295a.

31. Sghaier Chriki and Jean-François Hocquette, "The Myth of Cultured Meat: A Review," *Frontiers in Nutrition* 7 (February 2020): 7, doi:10.3389/fnut.2020.00007.

32. Matthias Stadtfeld and Konrad Hochedlinger, "Induced Pluripotency: History, Mechanisms, and Applications," *Genes & Development* 24, no. 20 (October 2010): 2239–2263, doi:10.1101/gad.1963910.

33. Naira Caroline Godoy Pieri et al., "Stem Cells on Regenerative and Reproductive Science in Domestic Animals," *Veterinary Research Communications* 43, no. 1 (February 2019): 7–16, doi:10.1007/s11259-019-9744-6.

34. Neal D. Amin and Sergiu P. Paşca, "From Stem Cells to Embryo Models in the Laboratory," *Nature* 610: 39–40; Cassandra Willyard, "Mouse Embryos Grown without Eggs or Sperm: Why, and What's Next?" *Nature* 609: 230–231.

35. Harunobu Kagawa et al., "Human Blastoids Model Blastocyst Development and Implantation," *Nature* 601, no. 7894 (January 2022): 600–605, doi:10.1038/s41586-021-04267-8; Susanne C. van den Brink and Alexander van Oudenaarden, "3D Gastruloids: A Novel Frontier in Stem Cell-Based in Vitro Modeling of Mammalian Gastrulation," *Trends in Cell Biology* 31, no. 9 (September 2021): 747–759, doi:10.1016/j.tcb.2021.06.007.

36. Kaoru Miyazaki et al., "Generation of Progesterone-Responsive Endometrial Stromal Fibroblasts from Human Induced Pluripotent Stem Cells: Role of the WNT/CTNNB1 Pathway," *Stem Cell Reports* 11, no. 5 (November 2018): 1136–1155, doi:10.1016/j.stemcr.2018.10.002.

37. Nidhi Subbaraman, "First Monkey–Human Embryos Reignite Debate Over Hybrid Animals," *Nature* 592, no. 7855 (April 2021): 497–497, doi:10.1038/d41586-021-01001-2.

38. Leonardo Beccari et al., "Multi-Axial Self-Organization Properties of Mouse Embryonic Stem Cells into Gastruloids," *Nature* 562, no. 7726 (October 2018): 272–276, doi:10.1038/s41586-018-0578-0; David Cyranoski, "Lab-Grown Cells Mimic Crucial Moment in Embryo Development," *Nature* 582, no. 7812 (June 2020): 325–325, doi:10.1038/d41586-020-01757-z; Nidhi Subbaraman, "Lab-Grown Structures Mimic Human Embryo's Earliest Stage Yet," *Nature* 591, no. 7851

(March 2021): 510–511, doi:10.1038/d41586-021-00695-8; Kagawa et al., "Human Blastoids Model Blastocyst Development and Implantation"; van den Brink and van Oudenaarden, "3D Gastruloids."

39. Toshio Takahashi, "Organoids for Drug Discovery and Personalized Medicine," *Annual Review of Pharmacology and Toxicology* 59, no. 1 (January 2019): 447–462, doi:10.1146/annurev-pharmtox-010818-021108; Jihoon Kim, Bon-Kyoung Koo, and Juergen A. Knoblich, "Human Organoids: Model Systems for Human Biology and Medicine," *Nature Reviews Molecular Cell Biology* 21, no. 10 (October 2020): 571–584, doi:10.1038/s41580-020-0259-3.

40. Heidi Ledford, "Scientists Grew Tiny Tear Glands in a Dish—Then Made Them Cry," *Nature* 591, no. 7851 (March 2021): 515–515, doi:10.1038/d41586-021-00681-0; Takahashi, "Organoids for Drug Discovery and Personalized Medicine."

41. Ledford, "Scientists Grew Tiny Tear Glands in a Dish—Then Made Them Cry"; Takahashi, "Organoids for Drug Discovery and Personalized Medicine."

42. Alexandra K. Eicher et al., "Functional Human Gastrointestinal Organoids Can Be Engineered from Three Primary Germ Layers Derived Separately from Pluripotent Stem Cells," *Cell Stem Cell* 29, no. 1 (January 2022): 36–51.e6, doi:10.1016/j.stem.2021.10.010.

43. Ross J. Porter, Graeme I. Murray, and Mairi H. McLean, "Current Concepts in Tumour-Derived Organoids," *British Journal of Cancer* 123, no. 8 (October 2020): 1209–1218, doi:10.1038/s41416-020-0993-5; Yoshimasa Saito et al., "Establishment of Patient-Derived Organoids and Drug Screening for Biliary Tract Carcinoma," *Cell Reports* 27, no. 4 (April 2019): 1265–1276.e4, doi:10.1016/j.celrep.2019.03.088.

44. Ashley Del Dosso et al., "Upgrading the Physiological Relevance of Human Brain Organoids," *Neuron* 107, no. 6 (September 2020): 1014–1028, doi:10.1016/j.neuron.2020.08.029; Aaron Gordon et al., "Long-Term Maturation of Human Cortical Organoids Matches Key Early Postnatal Transitions," *Nature Neuroscience* 24, no. 3 (March 2021): 331–342, doi:10.1038/s41593-021-00802-y.

45. Kevin G. Chen et al., "Pharmacological Analysis of CFTR Variants of Cystic Fibrosis Using Stem Cell-Derived Organoids," *Drug Discovery Today* 24, no. 11 (November 2019): 2126–2138, doi:10.1016/j.drudis.2019.05.029; Rahul Mittal et al., "Organ-on-Chip Models: Implications in Drug Discovery and Clinical Applications," *Journal of Cellular Physiology* 234, no. 6 (June 2019): 8352–8380, doi:10.1002/jcp.27729.

46. Fjodor A. Yousef Yengej et al., "Kidney Organoids and Tubuloids," *Cells* 9, no. 6 (May 2020): 1326, doi:10.3390/cells9061326.

47. Yousef Yengej et al., "Kidney Organoids and Tubuloids"; Tristan Frum and Jason R. Spence, "HPSC-Derived Organoids: Models of Human Development and Disease," *Journal of Molecular Medicine* 99, no. 4 (April 2021): 463–473, doi:10.1007/s00109-020-01969-w.

48. Aaron Cohen et al., "Mechanism and Reversal of Drug-Induced Nephrotoxicity on a Chip," *Science Translational Medicine* 13, no. 582 (February 2021): eabd6299, doi:10.1126/scitranslmed.abd6299; Dena Wimpfheimer, "Can Chips Replace Animal Testing?," *EurekaAlert!*, March 2021, https://www.eurekalert.org/news-releases/825907.

49. Fernanda Majolo et al., "Use of Induced Pluripotent Stem Cells (IPSCs) and Cerebral Organoids in Modeling the Congenital Infection and Neuropathogenesis Induced by Zika Virus," *Journal of Medical Virology* 91, no. 4 (April 2019): 525–532, doi:10.1002/jmv.25345.

50. Anand Ramani et al., "SARS-CoV-2 Targets Neurons of 3D Human Brain Organoids," *EMBO Journal* 39, no. 20 (October 2020): 1–14, doi:10.15252/embj.2020106230.

51. Stefano L. Giandomenico et al., "Cerebral Organoids at the Air–Liquid Interface Generate Diverse Nerve Tracts with Functional Output," *Nature Neuroscience* 22, no. 4 (April 2019): 669–679, doi:10.1038/s41593-019-0350-2.

52. Charlie Schmidt, "System in a Dish," *Nature* 597 (September 2021): S22–S23.

53. Ranmal A. Samarasinghe et al., "Identification of Neural Oscillations and Epileptiform Changes in Human Brain Organoids," *Nature Neuroscience* 24, no. 10 (October 2021): 1488–1500, doi:10.1038/s41593-021-00906-5.

54. Ryan Britt, "The Most Intellectual Dystopia of All Time: Woody Allen's Sleeper," *TOR.com*, April 2011, https://www.tor.com/2011/04/13/the-most-intellectual-dystopia-of-all-time-woody-allens-sleeper/.

55. Sara Reardon, "Can Lab-Grown Brains Become Conscious?," *Nature* 586, no. 7831 (October 2020): 658–661, doi:10.1038/d41586-020-02986-y; Elke Gabriel et al., "Human Brain Organoids Assemble Functionally Integrated Bilateral Optic Vesicles," *Cell Stem Cell* 28, no. 10 (October 2021): 1740–1757.e8, doi:10.1016/j.stem.2021.07.010.

56. Sabina Kanton et al., "Organoid Single-Cell Genomic Atlas Uncovers Human-Specific Features of Brain Development," *Nature* 574, no. 7778 (October 2019): 418–422, doi:10.1038/s41586-019-1654-9.

57. Iva Kelava et al., "Androgens Increase Excitatory Neurogenic Potential in Human Brain Organoids," *Nature* 602, no. 7895 (February 2022): 112–116, doi:10.1038/s41586-021-04330-4.

58. Ariana Remmel, "Neanderthal-like 'Mini-Brains' Created in Lab with CRISPR," *Nature* 590, no. 7846 (February 2021): 376–377, doi:10.1038/d41586-021-00388-2; Cleber A. Trujillo et al., "Reintroduction of the Archaic Variant of *NOVA1* in Cortical Organoids Alters Neurodevelopment," *Science* 371, no. 6530 (February 2021): eaax2537, doi:10.1126/science.aax2537.

7

I Want to Be Your Dog

Man, do not exalt yourself above the animals, they are without sin,
while you defile the earth by your appearance on it.
— *The Brothers Karamazov*, Fyodor Dostoevsky

Flying at an altitude of 36,000 feet, Choupette looks out of the window of her private jet, her clear blue eyes reflecting the color of the sky. The stewardess serves her lunch—a mixture of smoked salmon and caviar. Perhaps she is thinking about how fortunate she is? She is both extremely beautiful and extremely rich. With millions of Instagram followers, she is a marketing phenomenon selling huge numbers of books and fashion accessories to her many fans. In addition, following the recent death of the love of her life, she has inherited a fortune totaling hundreds of millions of dollars. This may be all the more remarkable considering that Choupette is a cat. A blue-eyed Birman with long soft silky white fur, Choupette has shared most of her life with the fashion designer Karl Lagerfeld, who made no secret of the fact that he loved her to distraction and wanted to marry her, and would have done so but for the fact that marrying "animals" was not countenanced by either the Catholic Church or the laws of his adopted country, France. Nevertheless, since they met in 2011, Choupette and Lagerfeld were constant companions living together, it seems, in perfect harmony. It is said that her influence over him was so great that he introduced fake animal fur into his last collections for Dior, Fendi, and his own eponymous label, and that after years of resisting the idea.[1]

Although Choupette's situation is certainly unusual, there is a long history of human devotion to animals. There is no doubt that animals can occupy places in our lives that are as close as those of any human. Celebrity stories of such relationships can be found daily in the media and throughout history and range from people like Barbara Streisand, who cloned her beloved dog Samantha, to Sir Isaac Newton, who used his great genius to devise the

The Rise and Fall of Animal Experimentation. Richard J. Miller, Oxford University Press. © Richard J. Miller 2023.
DOI: 10.1093/oso/9780197665756.003.0007

"cat-flap," allowing his pet cat Spithead access to his rooms in Trinity College, Cambridge, when he locked himself away from everybody else so that he could invent calculus, gravity, optics, or some other little scientific idea.[2]

Close mutual interactions between humans and animals in the form of pet keeping and other activities are increasing every year. According to the American Pet Products Association (APPA) more than two-thirds of US households owned pets in 2021, an increase from 56.8% in 2016. According to the APPA, millennials represent the largest segment of pet owners for all pet types, which, apart from cats and dogs, can also include birds, reptiles, small exotic mammals, and diverse kinds of fish.[3]

Spending on pets is skyrocketing and encompasses all aspects of a pet's life including veterinary care, gourmet food, pet accessories, and so on. Again, the APPA reported that overall spending in the US pet industry increased from $72.56 billion in 2018 to $103.6 billion in 2020, the first COVID-affected year—a colossal amount.[4] And it's not just in the United States where these numbers are growing; similar trends have been noted all over the world. In today's fragmented and COVID-infected world, where people live an increasingly isolated existence, a companion animal can certainly help to reduce loneliness and allow people to maintain some kind of contact with the living world. And animals don't only impact our lives when we are at home; they can help people connect with the world in general and are becoming more and more a part of our everyday existence. As well as just walking your dog, there are now dog parks to visit, cat cafes to hang out in, and other opportunities for appreciating how animals can become part of your life in general.

Animals help us get in touch not only with ourselves but also with others. In a recent experiment carried out in Paris, a man approached some 240 different young women in a park in an effort to obtain their phone numbers. He was successful just 9% of the time. But when he tried the experiment a second time, now accompanied by a dog, his success rate jumped to 28%—a gigantic increase.[5]

Animals help us feel comfortable with ourselves and with others both at home and at work and allow us to connect with a real living object rather than a cell phone or a computer. You not only love your dog or cat, but it seems clear that they love you as well, and that is extremely important. And although we don't all go to the same extremes as Karl Lagerfeld, people are trying to show their affection for their pets in increasing numbers of ways. Pets are no longer fed scraps from the table but are much more likely to eat an array of gourmet pet foods and have a comfortable bed to lie in. For example, you can purchase a large bag of Open Farm Pet's "Wild-Caught Salmon & Ancient Grains Dry Dog Food" for $72. It contains not only wild salmon but also steel-cut oats, quinoa, chia seeds, and "superfoods like coconut oil, pumpkin and turmeric." The increase in pet ownership is really part of a more general phenomenon where people, whose lives are increasingly physically and emotionally constrained by what they perceive to be enormous faceless governments and businesses interests, try to react by exhibiting greater respect for the environment and a connection with Nature: more organic food, less factory farming and meat production, more vegetarianism and veganism, more fake fur and leather, more solar energy, and so on.

It might be argued that the increase in pet ownership and general animal awareness is just the provenance of wealthy people in developed countries who have the time and money to indulge themselves in recreational pastimes, and that owning a pet is another such activity like going to the beach or playing tennis. If that were the case, however, then it would be predicted that pet ownership would not be as common among less affluent societies or classes. But nothing could be further from the truth. This is a mass movement. The numbers speak for themselves as people who share these opinions become a larger and larger part of society and demand to be heard not only by speaking out but also by flexing their credit cards.

The idea that close, emotionally rich relationships with animals are good for the general state of your health is clearly true. Back in the 1970s, experiments were carried out by Dr. Erika Friedmann, who looked at the impact of lifestyle and related factors on the survival of a large group of heart attack victims. Friedmann showed that pet owners were more likely to survive for one year after a heart attack than nonowners.[6] Subsequently, a huge number of investigations have confirmed effects of this type. Pet ownership has beneficial effects on many objective health measures including reducing heart rate, blood pressure, cholesterol, serum triglycerides, and levels of stress hormones such as cortisol, as well as simultaneously increasing levels of oxytocin, a hormone that seems to improve human sociability.[7] Pet owners seem to have generally more robust health and clearly make fewer visits to the doctor's office. It's really a two-way street—animals and humans can live together in harmony and can be good for one another.

As a wave of public opinion insists on greater respect for animals, where does this leave science? If people object to animals being abused and tortured or, if you will, "experimented upon," what are scientists to do? Are they going to become more transparent? Are they going to open up their laboratories to the general public and allow people to come in and see them give animals epileptic seizures, jam electrodes into their brains, burn them, maim them, and give them cancer? Good luck with that! If you ask scientists why they do these things, the most common answer you will receive is that, apart from their own personal curiosity, which is fair enough, they are serving the public interest, which demands cures for terrible diseases. This is their mantra. However, I doubt very much whether many members of the public are aware of exactly what is going on in the laboratories where scientists are making attempts to find these cures—attempts that are, more often than not, completely fruitless. What would people think if they had a clear idea of the biomedical scientific modus operandi? They would probably be horrified. The question as to what percentage of people are "against" animal research is difficult to answer precisely. It depends on what kind of research one is talking about, what kinds of animals are involved, and so on. Whereas just over half of individuals in the United States in 2018 said they were against the use of animals in research, this number rose to over 70% if the research in question was specifically for making cosmetics. What is certainly clear, however, is that the trends demonstrating opposition to animal research, whatever the category, are rising every year.[8]

I am speaking of course of science as it is currently practiced, the result of historical processes that we have already discussed. The thing is that science is an extraordinarily important cultural activity, and it can provide the answers that we need, including the cures for horrible diseases—but how it achieves these ends needs to change. Science will only change very slowly if left to its own devices. But, as we shall see, there are now increasing efforts to legislate what scientists can and can't do. These changes are being put into effect as a result not only of the demands of PETA or other animal welfare groups but also, perhaps much more importantly, of public opinion in general. This is a grassroots movement of massive proportions, a tsunami of love for animals, and it is demanding change. Moreover, at the same time, new scientific methods are making animal experiments increasingly obsolete, so we simply don't need to do them anymore. Using these new methods, the scientists of the future will be able to further our understanding of Nature and provide answers to medically important questions in a manner that is more commensurate with the overall goal of being a human being living in an animal-loving world. And in the end, it will be the weight of public opinion and the resulting legislation that will ultimately afford animals the respect they deserve under the law. If it is public opinion that eventually determines what scientists can do and how they do it, it is important for us to try and understand how our opinions concerning animals developed and where they may go in the future. Did we always keep pets and "companimals"?

How did the question of animal welfare become a mass movement, and how was it politicized? Of course, it is important to know and understand what historically important individuals like Aristotle, Galen, Augustine, Harvey, Descartes, and Boyle thought because they exerted a great deal of influence on science and how it goes about its business. But these were very unusual people operating far from the public eye. What, on the other hand, did normal people think about animals, and how did their attitudes develop? When did people begin to raise their voices en masse in support of animal welfare and against animal abuse? How has the voice of public opinion developed as a context in which science must operate? How has it affected science, and how will it do so in the future?

Ancient Pets

It seems clear that in antiquity, even in large cities, it was normal for people to live in close proximity to their animals. Domestic animals such as cows, goats, and pigs didn't just live in the countryside on farms but frequently lived together with humans, sharing their houses and living with them side by side.[9] Although this practice may have initially arisen because animals were useful as sheep herders, as guards, or to keep away rodents, from the very beginning it would have been common for humans to have responded to their animals with feelings of love and affection because of their natural empathy for other creatures and our assumptions about their mutual affection for us, something that, then as now, would have helped to support us emotionally. How companion animals really "feel" about their owners is, of course, unclear, and whether they "love" us is perhaps less important than the fact that we think they do. Animals, it seems to us, are guileless and accept us unconditionally, without criticism and without rejection.

The idea that animals could provide real companionship, beyond their general practical utility, appears to be extremely old. The earliest known domestic dog burial remains have been dated to around 10,000 BCE from a Natufian site in Israel where a four- to five-month-old puppy was found buried intact with a human skeleton, something that really speaks for itself.[10]

Somewhat more recently, humans buried with dogs have been found at sites in Germany and Israel, and a cat-human burial dating from about 9,500 years ago has been discovered on the Mediterranean island of Cyprus.[11] It seems reasonable to suppose that the decision to bury an animal with a human implies awareness that their close relationship would continue into the after-life. We know that the spirits of animals were held in religious reverence by the earliest societies. The Egyptians, for example, often depicted their gods in the form of animals—Anubis as a jackal or Bastet as a cat and many others (Figure 7.1). Indeed, the cat was an animal of special significance for the Egyptians. In tombs, cats are sometimes shown chasing birds and playing. Cats were useful and could kill rodents and snakes. In mortuaries, they were sometimes illustrated with a dagger, cutting through Apopis, the snake deity threatening Ra (the Sun) at night in the Underworld.

Figure 7.1 Statue of an ancient Egyptian cat.
Image from Shutterstock.

There are many examples of great care in the depiction of cats in Egyptian art, and there are numerous instances of cat mummies being discovered in Egyptian tombs. In 2018, a 4,500-year-old tomb was discovered in the Saqqara necropolis on the outskirts of Cairo containing dozens of mummified cats, 100 gilded wooden cat statues, and a bronze statue representing the goddess of cats, Bastet. Sometimes food has also been found to accompany these ritual burials, another indication that they were buried in expectation of an afterlife. It seems as though Egyptians, at least those who were wealthy enough, kept cats and other animals and had relationships with them that are not all that dissimilar to many pet owners today.

Given the fact that the culture of science as we know it really began in Greece, this seems like a good place to begin asking how humans and animals interacted in antiquity. Individual opinions about animals as recorded by famous writers in Greece and Rome certainly ran the gamut. For example, Aristotle placed animals below humans in the "great chain of being." But we also know that there were some important writers in antiquity who greatly valued the lives of animals. This was true, for example, of Pythagoras, Theophrastus, Plutarch, and Porphyry, and we know a great deal about the reasoning involved in their opinions (see Chapter 2). Pythagoras and Empedocles spoke of a kinship based on humans and animals being composed of the same elements and sharing the same unifying metempsychotic spirit that pervaded the entire universe, something that could lead to reincarnation as a different species. Empedocles stated, "For already have I once been a boy and a girl, and a bird and a dumb sea fish." The Neo-Platonist Porphyry, who lived during the early phase of Christianity in the third century CE, took issue with Stoic and Christian attitudes that animals existed only for use by humans, something that he thought made no sense. Was it also true that humans were specifically created so that they could be eaten by crocodiles? But this ultimately proved to be a minority view and gained little traction in the face of Christian hegemony.

There are certainly some well-known examples of celebrity pet keeping in ancient Greece. Perhaps the most famous of these are Bucephalus the horse and Peritas the dog, two famous companimals of Alexander the Great. It seems that Alexander was greatly attached to both these creatures, and a considerable folklore has grown up describing their heroic exploits. It wasn't all legend. Both animals clearly existed and Alexander's feelings for them were reflected by the fact that he founded cities named to celebrate their lives. Bucephalus, who had been Alexander's companion since his youth, died around the age of thirty during Alexander's invasion of northern India, and he founded the town of Bucephala in the Punjab to celebrate the life of his equine companion.[12] When Peritas died around the same time, he also founded another city in India named after his dog.[13] Alexander loved founding new cities, and he created around twenty of them. As he evidently thought rather a lot of himself, all these cities were called Alexandria. The fact that the cities of Bucephala and Peritas were called something else speaks to the great esteem in which Alexander held his two beloved companimals.

But was pet keeping in Greece just associated with the aristocracy, or did it extend to the *hoi polloi* as it does today? It is likely that some degree of pet keeping was widespread in ancient Greece and, to some extent, this transcended class. Dogs were the favorite type of pet, but other animals including cats, horses, sheep, goats, parrots, and even snakes could be the objects of human affection and, if they were sufficiently exotic, could also add a certain social status to their owner. The animal might be expensive and novel and have a similar effect as somebody owing a Ferrari or an Aston Martin today. Porphyry had a famous pet partridge; he wrote, "In Carthage we reared a partridge that came flying tamely to us. As time went by, and familiarity made it very tame, we saw it not only greeting us, attending us and playing with us, but even uttering in response to our utterances, and, as far as it could, answering us. And this it did not in the way that partridges usually call each other, but differently and not when one was silent. It was only when one uttered that it uttered in reply"

The prominent position of animals in ancient Greek society was strongly reflected in literature and art, where representations of animals have provided us with indications of how they were treated. As we have seen (see Chapter 2), depictions of animals are as old as art itself, and their images have been found portrayed in every medium for at least the last 30,000 years. People have "sculpted them, painted their pictures and, since the 1840s, taken their photographs."[14] From these sources it appears that the most common pet of all in antiquity was the lapdog known as the Melitan/ Maltese, a small, fluffy dog, commonly white in color, which was imported into Greece and Rome from Malta.[15] While it was not the only kind of miniature dog mentioned in surviving texts, it was by far the most common, and there are an enormous number of references to these animals in Greek and Roman art and literature. These little lapdogs, which were particularly associated with ladies, seemed to have rivaled their modern counterparts in their status as pampered pets. Juvenal, satirical as ever, wrote that sometimes women would gladly send their husbands to the grave to save the lives of their little dogs (Figure 7.2).

Nevertheless, amid the wide array of daily interactions with animals, their humane treatment or otherwise did not appear to be the subject of any kind of widespread debate as to whether animals had something akin to rights or whether laws should be passed to protect them. Rather, their treatment was more often than not the result of common practical considerations. Ancient peoples needed to be pragmatic, and kinder feelings may often have been forced aside by the realities of everyday life. Deliberate cruelty was reviled no doubt, but incidental suffering was presumably largely disregarded when it competed with human needs and desires. Such considerations suggest that attitudes toward animals varied a great deal in ancient Greece, as they do today, but never really coalesced into a popular movement designed to use political power in the defense of animal lives. It is certainly true that intellectuals like Aristotle or Plutarch[16] wrote a great deal about animals,

Figure 7.2 Greek vase with Melitan (Maltese) dog.
Image used with permission from the University of Pennsylvania Museum.

but whether there was a general public debate about them is another matter; there is no sign that there was.

Similar conclusions can be made with respect to other periods in antiquity such as during the Roman Empire. Indeed, although, as in Greece, there is evidence for pet keeping and close association with animals, a much more publicized attitude attributed to the Romans was the great cruelty they exhibited in circumstances associated with public events where animals (and humans) were tormented for the purpose of public entertainment. Indeed, as we try to understand the close emotional bonds that form between humans and animals, how do we simultaneously explain human behaviors such as hunting, using animals as sporting objects, as in ancient Rome, or bull, dog, bear, or cock fights throughout the ages? It would be useless to deny, as noted later by John Locke, that there are behaviors that come naturally to humans that are cruel and just as ancient as the altruistic and loving aspects of human behaviors exhibited toward animals and each other.

Medieval Attitudes

The long period of time between the fall of the Roman Empire and the coming of the Scientific Revolution in the seventeenth century was not represented by much progress in the sciences (see Chapter 3), and there was little to encourage new and enlightened attitudes toward animals. Galen was the model for medicine in both Christian Europe and the Arab empires, and his attitudes and methods were considered sacrosanct. Aristotle's charge that animals lacked reason together with Christian concepts of human superiority meant that there were few influential public figures who would speak out in the defense of animals during the Dark Ages. Animals were fine as long as they were kept in their "natural" position of servitude. After all, Christians were concerned with what the Bible said, and God had said to Adam and Eve, "Be fruitful and multiply, and fill the earth and subdue it; have dominion over the fish of the sea and over the birds of the air and over every living thing that moves upon the earth" (Genesis 1–28), and to Noah, "The fear of you and the dread of you shall be upon every beast of the earth, and upon every fowl of the air, upon all that moveth upon the earth, and upon all the fishes of the sea; into your hand are they delivered. Every moving thing that liveth shall be meat for you" providing a pretty clear road map for humanity's interactions with Nature.

In Europe in the Middle Ages, the general superiority of humans over animals was thought to be "natural," as reflected by things like the upright stature of humans: "All such animals are prone to the ground—because of the weight of their head and the earthy character of their body, they tend to bear themselves in a horizontal plane, their innate heat being inadequate to maintain them in an erect posture," wrote Albertus Magnus. Even as late as the seventeenth century John Donne wrote, "Wee attribute but one privilege and advantage to Man's body above other moving creatures, that he is not, as others, groveling, but of an erect, of an upright form, naturally built, and disposed to the contemplation of heaven." There were numerous other differences as well that were thought to reflect the superiority of humans, some quite bizarre, such as the fact that humans had less hair and could laugh, something that animals were seemingly unable to do.[17] All of this highlighted the long influence of Christian thought from the fourth century CE, when St. Ambrose and St. Augustine had distinguished "brute animals from rational humans," to Thomas Aquinas in the thirteenth century, whose attempts to use Aristotelean rationalism to redefine Christian thought yielded similar conclusions about animals. St. Augustine had declared that "animals are without intellect and are not made in the image of God" and thought that animals would not participate in the Resurrection or have an afterlife. And as the Thomists declared, "The life of animals is preserved not for them but for man." God's word, then, through the efforts of his Christian followers, condemned animals to their fates.

Even as late as the seventeenth century, Europeans simply couldn't understand views that didn't assign the whole of Nature to something they could use for any purpose that they saw fit. As discussed by Keith Thomas, "When travelers came back with reports of how Eastern Religions held a totally different view, and how Jains, Buddhists and Hindus respected the lives of animals, even insects, the general reaction was baffled contempt. It was 'unaccountable folly,' of the Hindus, thought one seventeenth century observer, to let a widow burn on her husband's pyre and yet be so sparing of the lives of insignificant creatures, as if the life of man were of less consequence and consideration than that of a beast."[18]

There was yet another reason intimate relationships with animals were looked down upon in medieval times, and that was the idea of witchcraft. In Europe in the sixteenth and seventeenth centuries, it was widely believed that witches made use of animals as "familiar spirits" who could carry out their commands for performing evil deeds on victims of their choice. These "familiars" were usually small animals, such as cats, dogs, mice, or toads. Because of this association, people suspected of witchcraft would not help their cause by displaying affection for a pet, and it is clear from the court records that evidence of pet ownership was commonly cited during witch trials.

Witches aside, in medieval times the precise relationships between the vast majority of common people and their animals probably hadn't changed that significantly since antiquity, and it was generally assumed that the purpose of animals was to serve humanity. People presumably did have close relationships with animals, especially dogs, although these would also be looked upon as sources of labor for taking care of herds of sheep and other animals. Cats were also kept as pets and would have to earn their keep by controlling rodents; moreover, they were tainted by their supposed role as witches' familiars. As in ancient times, common people lived in close proximity with their animals, often sharing their houses with them, and although they would not have had any qualms about slaughtering animals for food, they must often have had affectionate feelings for them. Genuine pet keeping without any other implications presumably only occurred, as in antiquity, in the case of wealthy or aristocratic individuals.

Some insights into the way humans viewed animals in the Middle Ages come from the precise values they put on them in material terms. "The Visigoths ranked a nobleman as 500 times more valuable than the noblest animal. The Franks valued a man's nose or ear at twice the value of a stallion. The Burgundians valued one tooth of a nobleman at more than twice the value of the most prized falcon. One tooth of someone from the lowest class equaled a valuable hunting dog. For the Alamans, the thumb of a man was almost twice as valuable as the best stallion."[19] In conclusion, human attitudes toward animals in medieval times were generally anthropocentric, driven by the dominance of Christian thinking. The fate of animals was not a matter for public debate. There were no great names writing or speaking out in favor of raising the status of animals in the world.

Eventually, beginning with the Renaissance and accelerating during the sixteenth and seventeenth centuries, the revolution in scientific thinking began to encourage changes in how humans perceived their position in the cosmos (see Chapter 3). Following Copernicus's revelation that Earth traveled through space by circling the sun, rather than the other way around, the position of humanity as the center of the universe began to shift and attitudes began to change, including attitudes toward animals. Over the next several hundred years a revolution in public thinking about animals took place that finally resulted in the grassroots movements opposing cruelty to animals that we see today. The Scientific Revolution was represented not only by thinkers like Descartes who thought of animals as being irrational quasi-clockwork mechanisms incapable of feeling real pain but also by the rise of thinking that opposed this trend and resulted in a new appreciation of animals as conscious entities on their own terms. Christian Europe had characterized animals by taking the point of view attributed to Protagoras in the fifth century BCE that "Man is the measure of all things." But as we have discussed animals need to be appreciated and understood for their own merits (see Chapter 5).[20] A dog's behavior needs to be understood in terms of "what it is like to be a dog" and what is good or bad for a dog, not just as some kind of insipid reflection of human behavior so that it is inevitably judged as being inferior. It isn't inferior, just different. But who would speak out and raise the issue that animals should be understood according to their own merits?

The Rise of Modern Attitudes in Support of Animals

The man who initiated this kind of discussion in the modern world was the great sixteenth-century French philosopher Michel de Montaigne. Montaigne is mostly remembered for his writings that often took the form of "essays," a literary form he is credited with inventing, and which allowed him to comment on any subject he chose, often from a personal point of view.[21]

Montaigne's essays are of tremendous importance and his influence can be readily discerned in the works of later great writers including the likes of Francis Bacon and Thomas Browne in the next century, continuing right up to the present day. Montaigne himself was influenced by the philosophical outlook of Pyrrho of Elis, a Greek philosopher who founded a form of critical thinking known as "skepticism." Indeed, Montaigne was famous for his mantra "*Que sais-je?*" ("*What do I know?*").

Montaigne had a strikingly modern view of animals and how they should be treated, and his views will resonate powerfully with the millions throughout the world today who deplore their cruel treatment.[22]

Most of Montaigne's opinions about the treatment of animals appeared in his longest essay *The Apology for Raimond Sebond* (1580) and the preceding essay *On Cruelty*. Montaigne completely abandoned the view that human beings are something unique when compared to animals. Humans were just another kind of animal who had their own point of view. Toward the beginning of the series of statements on animals in the *Apology*, he wrote as follows on interactions with an animal he was very familiar with: "When I play with my cat, who knows if I am not a pastime to her more than she is to me?" And this is surely the case. Montaigne wrote:

> We have some tolerable apprehension of their meaning, and so have beasts of ours—much about the same. They caress us, threaten us, and beg of us, and we do the same to them . . . for what is this faculty we observe in them, of complaining, rejoicing, calling to one another for succour, and inviting each other to love, which they do with the voice, other than speech? And why should they not speak to one another? They speak to us, and we to them. In how many several sorts of ways do we speak to our dogs, and they answer us? We converse with them in another sort of language, and use other appellations, than we do with birds, hogs, oxen, horses, and alter the idiom according to the kind.

Indeed, Montaigne's view of the situation would lead to one of its obvious conclusions, as voiced some years later by Jonathan Swift when describing the adventures of Lemuel Gulliver, as to whether humans or animals were more "intelligent": who were the Yahoos and who were the Houyhnhms?

Why should it be asked Montaigne that we consider our actions to be the results of reasoned consideration and that we say what animals do is merely the result of instinct?

> So, I say, to return to my subject, that there is no apparent reason to judge that the beasts do by natural and obligatory instinct the same things that we do by our choice and cleverness. We must infer from like results to like faculties, and consequently confess that in the same reason, this same method that we have for working, is also that of the animals. Why do we imagine in them this compulsion of nature, we who feel no similar effect?

Moreover, the vaunted use of language by humans was not so unique:

> In one kind of barking of a dog, the horse knows there is anger, of another sort of bark he is not afraid. Even in the very beasts that have no voice at all, we easily conclude, from the society of offices we observe amongst them, some other sort of communication: their very motions discover it: As

infants who, for want of words, devise Expressive motions with their hands and eyes.

Montaigne wrote about the affective empathic response we have toward animals and their suffering and thought that it was this, rather than some argument based on rationality, that should govern our responses toward them. Hence, Montaigne concluded: "We owe justice to men, and mercy and kindness to other creatures that may be capable of receiving it. There is some relationship between them and us, and some mutual obligation." And Montaigne went further in his opinion of humanity, "having learned by experience, from the cruelty of some Christians, that there is no beast in the world to be feared by man as man." Montaigne's views, radical at the time, were to presage those of the French *philosophes* at the end of the eighteenth century who believed that humans and animals were linked together in a web of mutual respect and that animals should not be exploited for any anthropocentric reason.

Descartes, whose views of the matter were quite different from those of Montaigne, made reference to them in a 1646 letter to Margaret Cavendish, the Marquess of Newcastle, in which he stated his disagreement with Montaigne's attribution of "understanding or thought to animals," writing, "I cannot share the opinion of Montaigne and others who attribute understanding or thought to animals. I am not worried that people say that human beings have absolute dominion over all the other animals."[23] In both this letter and the *Discours de la methode* (1637), Descartes treated Montaigne's views as a completely unacceptable philosophical position. And of course, from the point of view of the basic experimental sciences, it has always been Descartes's way of thinking that has held sway.

The disparate nature of Descartes's and Montaigne's opinions on the abilities of animals really set the scene for the different types of scientific approaches in the use of animals that have echoed down through the years. Descartes's approach led to the current use of animals as models of humans that can be experimented upon in any way that is supposed to serve the needs of humankind, with few considerations that the animals involved are of value in themselves; this is opposed by other approaches, more associated with natural history and ecology, in which animals are studied as part of the general environment on more or less their own terms.

Although Montaigne's views on the relationship between humans and animals may seem strikingly modern and could have been articulated by many people today who feel great empathy toward animals, in the sixteenth century his thinking was not influential enough to create a public forum for debate concerning these matters. That would come later. In the seventeenth and eighteenth centuries both science and theology still had the goal of helping humanity establish complete domination over Nature and restore to humans what had been lost after the Fall in an attempt to elevate them once more to God's original image. Humans were viewed as unique and completely different from animals, which existed solely to serve them.

It is clear that the type of thinking originated by Descartes and his followers was extremely influential in shaping the course of science in the seventeenth century and beyond. Nevertheless, Cartesian ideas were not uniformly accepted throughout the continent and soon some inkling of an opposition to Cartesianism began to arise. In England, for example, which, as we shall see, would originate the popular backlash against science-associated cruelty to animals, Descartes's ideas were not as widely accepted as on the continent, and English writers of great importance such as John Locke thought that Cartesian notions about animal behavior "were against all evidence of sense and reason." Locke advocated the keeping of pets to encourage children to develop empathy and a sense of responsibility toward others. Locke wrote an essay entitled *Some Thoughts Concerning Education and Cruelty*, where he detailed his feelings on this matter.[24] Here Locke described the idea that mistreatment of animals by humans would encourage savage behavior between humans themselves. This is clearly one strand in the argument for not being cruel to animals. Locke also recognized that "the child is father to the man" and both empathic and cruel tendencies are present in humans at birth, and the development of the former and suppression of the latter is something that should be encouraged by cultures that consider themselves enlightened.

Of course, it wasn't only Locke who had opinions that didn't support all the implications of Cartesianism. At some point Descartes's theories, if viewed without modification, started to look old fashioned; Jonathan Swift represented the view held by many reasonable people "that there is nothing so extravagant and irrational which some philosophers have not maintained for truth." Some Enlightenment thinking began to encourage opinions that animals were more than just machines; David Hume, for example, thought it was obvious "that the beasts are endow'd with thought and reason as well as men."[25] As the Enlightenment advanced, it began to change people's attitudes toward Nature in general, leading to a more inclusive view of humans and animals. Natural historians like Gilbert White who were busy measuring and classifying Nature encouraged the objective view that we should appreciate all of Nature's animals for their own sake and not just because of their value or otherwise to humans. Hence, an animal or "brute" could not be ugly just because its features would have looked so if exhibited by a human. "If a horse be beautiful in its kin, and a dog in his, why should not the beetle be so in its kind? Unless we measure the form of all things by our own, that what is not like us must be held to be ugly," wrote Thomas Muffett. Naturally, Thomas Browne also had something to say on the matter: "I cannot tell by what logic we call a toad, a bear or an elephant ugly."

During the course of the seventeenth century women also began to provide a novel sensibility that would greatly influence attitudes toward animals. The very first female writers who actually dared to publish books under their own names began to appear. A particularly remarkable instance of this was the work of "Mad Madge," Margaret Cavendish, Duchess of Newcastle, who wrote an enormous number of books of poetry, fiction (including the first science fiction novel), nonfiction, and plays. Cavendish was regarded as something of an eccentric who was famous for not only her controversial opinions but also her personal style of dress—she was the Vivienne Westwood of her time. Moreover, she was the first woman to be invited to a meeting of the Royal Society and was not scared to go head to head in discussions with the likes of Sir Robert Boyle. Cavendish's view of animals reflected those of Montaigne. She despised cruelty to animals as expressed by activities such as hunting, writing in her poem "The Hunting of the Hare" about hunters:[26]

> Men hooping loud, such acclimations make,
> As if the Devil they did prisoner take,
> When they do but a shiftless creature kill
> To hunt there needs no valient soldiers's skill.
> . . . for sport or recreation's sake
> Destroy those lives that God saw good to make:
> Making their stomachs graves, which full they fill
> With murthered bodies that in sport they kill.
> Yet man doth think himself so gentle mild,
> When he of creatures in most cruel wild.

In order for the views of someone like Locke or others sympathetic to animals to become more of a *cause célèbre* for the general public, they needed to be popularized. In this regard one of the most remarkable and widely discussed examples of the growing discomfort of common people to the mistreatment of animals, or at least the potential that they might think about these things seriously, was provided by the great artist and social critic William Hogarth. Hogarth was always ready to use the street life of London as a background for social commentary, and with *The Four Stages of Cruelty* (1751) he took aim at the inhumane treatment of animals and how this affected the character of humans.[27] This work was a set of four prints that were designed to be mass produced and widely distributed, basically for educational purposes (Figure 7.3). Against a background of everyday London life that many could identify with, the prints display the progress of Tom Nero, from his childhood when he abuses animals, to adulthood when he murders a woman and is convicted, following which his cadaver is used by medics for dissection practice. In the first print of the series, we observe numerous acts of barbarism, such as a boy throwing a cat out of the window, another putting out the eyes of a bird with a needle, and young Tom Nero sticking an arrow into a dog. In the second panel Nero has become a coach driver who abuses his horse, in the third he murders a woman, and in the fourth he is shown being dissected by scientists following his execution. This practice had become legal in 1752 when a law was passed allowing executed criminals to be used for medical demonstrations.[28]

Although it is clear that one of the messages of Hogarth's work was that human cruelty to animals encourages humans to be cruel to each other, the blatant nature of what he displayed certainly encouraged the viewer to pity the animals themselves on their own terms, something that Hogarth wrote was one of his primary intentions. Hogarth's thinking was not an isolated example. It was a reflection of trends that were springing up all over Europe, as in France, which, as we have seen, played a central role in the debate about animals and their treatment by scientists. Now numerous thinkers weighed in on the matter, including the French *philosophes*, influential Enlightenment thinkers, with Hogarth's contemporary Rousseau stating, "It appears, in fact, that if I am bound to do no injury to my fellow creatures, this is less because they are rational than because they are sentient beings: and this quality being common both to men and beasts, ought to entitle the latter at least to the privilege of not being wantonly ill-treated by the former."[29]

The rise of the middle classes and associated trends such as coffee house culture in the eighteenth century meant that themes that typified Enlightenment thinking, including matters concerning science and the details of scientific experimentation, started to be discussed much more

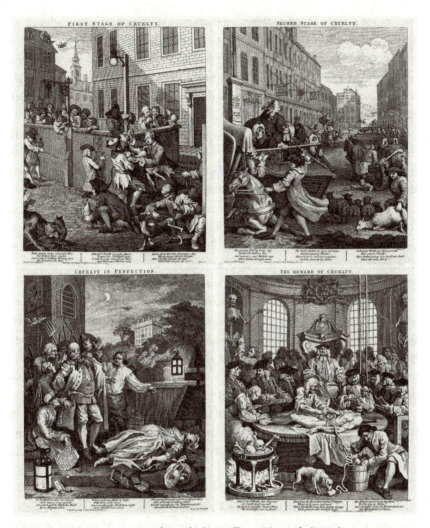

Figure 7.3 *The Four Stages of Cruelty* by William Hogarth (1751).
Image from Alamy.

broadly by the population at large and were powerfully depicted in art by
people like Joseph Wright of Derby (see Chapter 3). Science had become a
middle-class pastime. Condensed versions of the *Philosophical Transactions
of the Royal Society* started to circulate as well as magazines that enabled scientific matters to be widely commented on, something that was still possible
because technical scientific language had not yet become an impenetrable

barrier for debate among interested members of the public. And so, like all arguments, radically different opinions began to be heard.

The path of science that led from people like Vesalius and Harvey to the likes of Magendie and Bernard in the nineteenth century can be clearly traced, particularly with respect to their use of animals and vivisection. Backed by the forces of Christian thinking and Cartesian ideas, scientists performing these experiments could always justify what they were doing as being the natural order of things resulting from the God-given place of man in the universe. But in the face of some Enlightenment and post-Enlightenment thinking, these justifying attitudes began to be questioned, not only by Hogarth but also by others. The French Revolution encouraged the attitude that there was equality between men and furthermore between men and other creatures. In spite of the progress of science, a negative view of scientists was beginning to appear that is still with us today—that they spend too much time with their heads in the clouds doing things that are divorced from common sense. Jonathan Swift described Lemuel Gulliver, who, when visiting the flying island of Laputa, found that the "virtuosi" (a seventeenth-century word for a scientist)[30] were so unworldly that they required "flappers" to slap them in the face periodically and snap them from their perpetual reverie. And as far as research on animals was concerned, Joseph Addison (1710) criticized the virtuosi for becoming too preoccupied with the minutiae of vivisection: "They are so little versed in the world, that they scarce know a horse from an ox; but at the same time will tell you with a great deal of gravity that a flea is a rhinoceros and a snail an hermaphrodite."[31]

In the *Basset Table* (1705), author Susanna Centlivre (1669–1723) created a female scientist or virtuosa named Valeria who shared the same unfortunate passion for vivisection as her male compatriots.[32] At one point in the play Valeria reveals to her cousin Lady Reveller that she has killed and dissected her pet dove for thoroughly twenty-first-century reasons:

LADY: Oh, barbarous! killed your pretty Dove.

VAL: Kill'd it! Why, what did you imagine I bred it up for? Can Animals, Insects, or Reptiles, be put to a nobler Use than to improve our Knowledge? Cousin, I'll give you this Jewel for your Italian Greyhound.

LADY: What to cut to pieces? Oh, horrid! he had need be a Soldier that ventures on you; for my Part, I should dream of nothing but Incision, Dissection, and Amputation, and always fancy the Knife at my Throat.

Centlivre's contemporary Francis Coventry (1725–1754) wrote *The History of Pompey the Little or the Life and Adventures of a Lap-Dog,* which described its hero's adventures in escaping a vivisectional fate and raised the issue of the use of animals for testing drugs, which he viewed negatively:

A Dog might have been the emblematic animal of Esculapius or Apollo, with as much propriety as he was of Mercury; for no creatures I believe have been of more eminent service to the healing tribe than dogs. Incredible is the number of these animals, who have been sacrificed from time to time at the shrines of physic and surgery. Lectures of anatomy subsist by their destruction; Ward (says Mr. Pope) tried his drop on puppies and the poor; and in general, all new medicines and experiments of a doubtful nature are sure to be made in the first place on the bodies of these unfortunate animals.[33]

These are examples of the numerous eighteenth-century texts that had begun to discuss animals and their treatment in an altogether more sympathetic light, so different from the atmosphere that pervaded during the previous centuries. Now many extremely important thinkers and writers began lending their voices to the cause. The great poet Alexander Pope (1688–1744), a committed defender of animal rights, was completely against vivisection. Pope was the owner of four different dogs during his lifetime, all of which he named Bounce, and of which he was extremely fond. He had their portraits painted and commemorated them in verse. Interestingly, at one point Pope became the neighbor of the Reverend Stephen Hales, who was one of the most prominent vivisectors of the day and who made considerable contributions to the science of hemodynamics. They seem to have been on very good personal terms although Pope had a dim view of his neighbor's dealings with animals. As his friend Joseph Spence relates in the following exchange:

SPENCE: I shall be very glad to see Dr. Hales, and always love to see him; he is so worthy and good a man.
POPE: Yes he is a very good man, only—I'm sorry—he has his hands imbrued with Blood.
SPENCE: What, he cuts up rats?
POPE: Aye, and dogs too! (and with what emphasis and concern he spoke it.) Indeed, he commits most of these barbarities with the thought of its being of use to man. But how do we know that we have a right to kill creatures

that we are so little above as dogs, for our curiosity, or even for some use to us?[34]

In 1713 Pope voiced his opinions in an essay he contributed to Steele's periodical *The Guardian* entitled "Against Barbarity to Animals." And, of course, Pope is well known for his dictum "Know then thyself, presume not God to scan; The proper study of Mankind is Man." Pope was right about that, of course, as we would all do well to remember.

Indeed, by Pope's time the majority of English poets, essayists, and journalists were rejecting Descartes's concept of the "beast machine" and its implications for science in favor of something more empathic. The great Samuel Johnson, who had a lot to say about everything, made his opinions abundantly clear: "Among the inferior Professors of medical knowledge, is a race of wretches, whose lives are only varied by varieties of cruelty . . . [and] [i]t is time that a universal resentment should arise against those horrid operations, which tend to hard the heart and make the physicians more dreadful than the gout or the stone. Men who have practiced tortures on animals without pity, relating them without shame, how can they still hold their heads among other human beings."[35]

Comments like these are the types of things that might be said by many animal lovers today, and by the late eighteenth century opposition to the abuse of animals for research purposes was being widely discussed. Johnson's great friend and contemporary Sir Joshua Reynolds and many other artists of the time also began to highlight the connection between humans and animals, with pets becoming frequent additions to portraits of aristocratic families, the subtext being that they too were truly members of the family (Figure 7.4).

Philosophers of the time had mixed views concerning the place of animals in the world. The German idealist Immanuel Kant followed Aquinas's thinking that cruelty to animals resulted in cruelty to other humans and even used Hogarth's prints as illustrative examples of this tendency—but apart from that he had no issue with the idea that animals might be used by humans for any purpose including vivisection (see Chapter 8). On the other hand, an increasing number of people began to follow Montaigne's lead. Of particular importance in this regard was the utilitarian philosopher Jeremy Bentham (1748–1832). As we have already seen, in his *Introduction to the Principles of Morals and Legislation* Bentham asked the key question on behalf of animals: "Can they suffer?" Bentham prophesied that ultimately

Figure 7.4 *Miss Jane Bowles* by Sir Joshua
Reynolds (1775).
Image from Alamy.

"the day may come when the rest of the animal creation may acquire those
rights which never could have been withholden from them but by the hand
of tyranny."

Bentham's opinions about the status of animals became a great rallying cry
for the modern antivivisectionist cause. He really provided the first modern
critique of Christian thinking about man's relationship toward animals and
the idea that animals existed purely to be used by humans. After Bentham,
the issue of cruelty to animals was no longer solely seen from an anthropocentric point of view, but also from a theriocentric one, which considered
the protection of animals for their own sake. Bentham jettisoned the old
criterion of whether or not animals had rational souls and replaced it with
their capacity to suffer pain—again, something that would resonate with the
grassroots movement in support of animals today. Cruelty to animals is what
bothers people nowadays, not issues as to whether animals are rational or

not. And opinions like Bentham's were not unique to the British Isles. A similar point of view was also voiced by the influential German philosopher Arthur Schopenhauer (1788–1860), who strongly criticized Christian views regarding the worth of animals (see Chapter 8).

Philosophy aside, most people believed that it was important to live one's life according to the dictates of Christian faith, and we have seen what traditional Christian doctrine generally had to say about the use of animals. However, new religious attitudes began to take hold by the middle of the eighteenth century that had a profound influence on the relationship between science and Nature and asked the question, what really is the relationship between God and the natural world? Here an enormous change in emphasis was underway. Published in 1677, the *Ethics* of the Jewish Dutch philosopher Baruch Spinoza cast God in a completely different light. God for Spinoza was actually identical with the dynamic aspects of Nature, with him suggesting, "Deus sive Natura" (God or Nature)—that God and Nature were synonymous. By this Spinoza meant God was *Natura Naturans*—"a dynamic Nature in action, growing and changing, not a passive or static thing." Spinoza put forward the radical view that God and Nature were not purposeful and were indifferent to the fate of humanity. A view of Nature began to emerge where things were valid for their own properties and not just as a reflection of humans, thereby moving human attitudes back to a time prior to Aristotle and to the atomic theories of Democritus when Nature was an independent entity playing by its own rules and independent of the will of the gods. Moreover, if God and Nature were the same thing, then perhaps it followed that to constrain Nature by measuring it and manipulating it was to do the same to God. And this was an influential idea for many who were at the vanguard of the new culture of Romanticism, the coming worldview that was in many respects a reaction to the materialistic views of the Scientific Revolution. William Blake, one of the high priests of early Romanticism, was completely out of sympathy with the views of his personal trio of Enlightenment devils, Bacon, Newton, and Locke. Blake had little good to say about the scientific method, writing, "I will not reason and compare: my business is to create." People like Newton, thought Blake, who spent all their time measuring things and trying to cast Nature in a mechanical light just didn't get it. Blake painted Newton existing in a cave measuring the world with a compass. Like the prisoners in the cave in Plato's *Republic,* Blake thought Newton could only see the shadows of reality (Figure 7.5). Blake wrote:

Figure 7.5 *Newton* by William Blake (1795).
Image from Alamy.

O Divine Spirit sustain me on thy wings! That
I may awake Albion from His long & cold repose.
For Bacon & Newton sheath'd in dismal steel, their terrors hang
Like iron scourges over Albion, Reasonings like vast Serpents
Infold around my limbs, bruising my minute articulations
I turn my eyes to the Schools & Universities of Europe
And there behold the Loom of Locke whose Woof rages dire
Washd by the Water-wheels of Newton. black the cloth
In heavy wreathes folds over every Nation; cruel Works
Of many Wheels I View, wheel without wheel, with cogs tyrannic
Moving by compulsion each other.

Spinoza's pantheist thinking influenced many of the important figures of
the Romantic movement. In England Shelley and Coleridge paid close attention to his ideas, and in Germany so did the Romantic poets such as Novalis
and particularly Goethe, whose influence at the time was immense. It should
not be forgotten that in addition to being a writer, Goethe was also a scientist

manque. Goethe contributed to the idea of "Romantic biology" and, like Blake, thought that Nature was best studied by immersing oneself in it and appreciating and respecting it rather than deconstructing it, experimenting with it, and categorizing it. To this end Goethe developed an entire theory of color that he offered as an alternative to Newton's ideas suggesting that color vision was more dependent on the individual's ability to appreciate different colors rather than their actual existence as distinct physical entities. Goethe's theories appealed, in particular, to other Romantic artists such as the great painter Turner, who specifically referenced them in some of his most striking paintings (Figure 7.6).

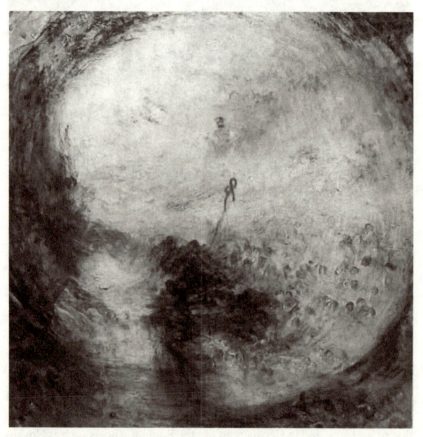

Figure 7.6 *Light and Colour (Goethe's Theory)—The Morning after the Deluge— Moses Writing the Book of Genesis,* J. M. W. Turner (1843).
Image from Alamy.

Romanticism encouraged the view that the reductionist science that had typified the Scientific Revolution involved an abuse of Nature. True science would involve a respect for Nature and so, by implication, a respect for all living creatures. This reflected the argument, associated with Jean-Jacques Rousseau, about the superiority of "natural" states and behaviors over those that were cultivated or artificial.

Edgar Allen Poe captured these ideas in his sonnet "To Science":

> Science! true daughter of Old Time thou art!
> Who alterest all things with thy peering eyes.
> Why preyest thou thus upon the poet's heart,
> Vulture, whose wings are dull realities?
> How should he love thee? or how deem thee wise,
> Who wouldst not leave him in his wandering
> To seek for treasure in the jewelled skies,
> Albeit he soared with an undaunted wing?
> Hast thou not dragged Diana from her car,
> And driven the Hamadryad from the wood
> To seek a shelter in some happier star?
> Hast thou not torn the Naiad from her flood,
> The Elfin from the green grass, and from me
> The summer dream beneath the tamarind tree?

And writers like Mary Shelley in her *Frankenstein or the Modern Prometheus* famously described the consequences of human interference with the normal course of Nature without regard for its consequences.

Overall, the dividing line between human beings and animals was becoming more porous once again and no more important, perhaps, than differences between different kinds of humans. All creatures were bound together through the notion of sympathy or compassion. A new continuum was starting to replace the classical and medieval concept of the "great chain of being" by adopting a nonhierarchical view and removing God from the equation.

These views were soon shown to reflect scientific reality with new ideas concerning evolution and development. Darwin's theories made the notion that there were absolute differences between man and other animals untenable. As the American philosopher James Rachels put it, "Darwinism undermined the traditional idea that human life has a special or unique worth." And the idea of a continuum didn't just apply to the evolution of animals. Parallel thinking about the manner in which embryos developed into adult animals produced similar conclusions. As we have seen (see Chapter 6), Darwin's disciple and popularizer, the German biologist Ernst Haeckel, suggested that animals developed from a single cell type, the *Stàmzelle*, or stem cell, itself a product of primordial spontaneously generated protein and mucus—the protoplasmic *monera*, a substance full of self-generating potential. Like a Darwinian tree of evolution, Haeckel could draw a tree of development illustrating how a stem cell could give rise to all the different tissues of an animal. The two processes, development and evolution, were analogous and could be connected according to a biogenetic law stating that "ontogeny recapitulates phylogeny."[36] This is a quintessentially Romantic idea suggesting the essential oneness of all aspects of Nature—man and animals were cast from the same mold, from the swirling inchoate mass that characterizes Turner's Goethe-inspired painting with reason, in the form of Moses, crystalized at the center (Figure 7.6). Just like the idea that all humans were equal and none were slaves, the parallel idea that animals and humans were all the result of the same generative biological processes indicated their closeness and discouraged the notion that animals merely existed to be used by humans. Indeed, it was the antislavery campaigner William Wilberforce who was responsible for some of the first attempts to pass laws protecting animals. And in the end, however incorrect many of its theories were as actual science, the idea of a unified Romantic biology would give rise to ideas and methods that will ultimately free scientists from the use of animals forever (see Chapter 6).

The Rise of Antivivisection and Animal Welfare

In the nineteenth century sentiments of this type were starting to provide the groundswell of public opinion that would finally result in the protection of animals under civil law. It was clearly the case that Hogarth's work was intended to stir debate by the man in the street about the effects of cruelty to animals, but where would this lead? Normally, if things stir the public sufficiently, they are reflected by changes in government and legislation. And eventually that is what transpired. Gradually, naturalists, educators, politicians, and campaigners rejected the old view that animals had been created solely to serve humankind. The time had come to take action. As soon as the Napoleonic Wars were over, great advances began to be made toward putting the ideas of the humanitarian writers of the eighteenth century into effect.

Britain took the lead in these reforms.[37] The reasons for this were doubtless complex, but the role of certain influences seems clear enough. One of these was the coming of the Industrial Revolution, which originated in Britain. The rapid urbanization of Britain had an enormous impact on how the general populace lived, moving from a predominantly agrarian way of life, which depended on farming and the countryside, to that of city dwellers and factory workers. This was accompanied by a growing middle class who had a more secure economic status and so more freedom to spend time on "civilized" aspects of life including what were thought to be "good causes," such as those that attacked cruelty to other humans or animals. Furthermore, as we shall see, England, as opposed to France, remained a monarchy, and the fact that Queen Victoria wholeheartedly endorsed the cause of animal welfare had enormous influence.

In 1800 and 1802 there were two initial attempts in Parliament to ban bullbaiting, led by Sir William Pulteney and supported by a variety of individuals including the antislavery reformer William Wilberforce as well as the playwright Richard Brinsley Sheridan; this was the first attempt anywhere in the world to introduce a law aimed at animal protection.[38] The horrible "sport" of bullbaiting was going out of fashion by this time and was regarded as low-hanging fruit for animal advocates to begin their reforming program through legislation. Nevertheless, both attempts were widely ridiculed in Parliament and defeated. In spite of this, the sentiment behind them did not disappear, and in 1809 a more far-reaching bill that sought to prevent any "wanton cruelty to animals" was introduced by Thomas Erskine, one of the early champions of animal welfare. Erskine's bill, introduced into the House of Lords, was the first to be presented in any Western legislature in an attempt to institute legal penalties for cruelty to animals in general. As Erskine put it:

> For every animal which comes in contact with man, and whose powers, qualities, and instincts are obviously constructed for his use, nature has taken the same care to provide, and as carefully and bountifully as for man himself, organs and feelings for its own enjoyment and happiness. Almost every sense bestowed upon man is equally bestowed upon them; seeing, hearing, feeling, thinking; the sense of pain and pleasure; the

passions of love and anger; sensibility to kindness, and pangs from un-kindness and neglect, are inseparable characteristics of their natures as much as our own.

Erskine's bill was also defeated. But now this defeat began to resonate with the public and to raise the ire of public opinion, with a popular magazine writing, "Surely few subjects in the whole compass of moral discussion can be greater than the unnecessary cruelty of man to animals which administer to his pleasure, his consolation and to the very support of his life!" But the op-position to the idea that animals should be protected by law was very deeply established, led by conservative elements in Parliament who didn't like the idea of a ban on hunting. Ultimately, it wasn't until July 22, 1822, that the first law in the world was passed that afforded some protection to animals, driven through Parliament by Erskine and another early reformer, Richard "Humanity Dick" Martin. This was "an act to prevent the cruel and improper treatment of cattle." The law proved to be no mere placeholder, and several prosecutions occurred in subsequent years. It was a historic breakthrough and the idea that cruelty to animals should be illegal began to gain real trac-tion. Further attempts at legislating against animal cruelty were passed, in-cluding the Cruelty to Animals Act of 1849, which provided protection for animals in general. The act was amended in 1876 and then replaced by the Protection of Animals Act in 1911. Other countries followed suit in passing legislation for the protection of animals. For example, in 1822, the courts in New York ruled that wanton cruelty to animals was a misdemeanor. Similar laws were passed in European countries such as France and by some other states in the United States.

It was not only in Parliament where new attitudes toward animals were apparent. The first inklings of a genuine grassroots movement were also beginning to stir, although initially these were led by "persons of quality" rather than the man in the street. In 1824, a group of upper-class individuals from a variety of different backgrounds held a meeting in a London coffee house, which led to the founding of the Society for the Prevention of Cruelty to Animals (SPCA)—a truly landmark event in the popular reaction to an-imal cruelty. The society pursued its aims vigorously but in 1840 was pro-vided with an enormous boost to its influence when Queen Victoria became its patron and its name changed to the Royal Society for the Prevention of Cruelty to Animals (RSPCA), a society that remains extremely active to this day. And the queen would continue to act, insofar as the law allowed her to, in support of the cause. In 1881, for example, her secretary wrote on her be-half to Prime Minister Gladstone, saying: "The Queen has seen with pleasure that Mr. Gladstone takes an interest in the dreadful subject of vivisection, in which she has done all she could, and she earnestly hopes that Mr. Gladstone will take an opportunity of speaking strongly against a practice which is a disgrace to humanity and Christianity."[39]

The association with Queen Victoria made the society and its message fashionable with the middle classes, for whom the overall pursuit of respect-ability was a fundamental aim. Although the society has had considerable influence throughout the world since its founding, at least initially it had a somewhat schizophrenic existence. Because its founding members came from the upper classes, which, in Britain at any rate, were closely associated with fox hunting and similar sports, many members of the RSPCA found that pursuits such as these were sacrosanct and outside the purview of its influence. But the writing was on the wall and others would take up that par-ticular challenge.

Indeed, it was the restricted reaction of the RSPCA to further legislation aimed at tackling problems like fox hunting that ultimately provided the stimulus for more radical factions to form their own society for fighting an-imal abuse, and this would have particular importance for science, because it was scientific methodology that was the focus of their attack. Vivisection in the service of physiology was pioneered by the likes of Harvey and his followers in the seventeenth century, culminating in the work and influence of Magendie and Bernard in France in the middle of the nineteenth century (see Chapter 3). Powerful images began to emerge illustrating what was going on in France that greatly influenced the British public. Now voices arose in England aimed at confronting science with respect to the practice of vivi-section, led by the journalist and humanitarian Frances Power Cobbe, who had been active in other causes including women's rights. Cobbe organized a group of like-minded people, many from the aristocracy as well as profes-sional people, and formed what was initially called the Victoria Street Society and then the National Anti-Vivisection Society, in an effort to persuade the government to pass legislation curbing the use of vivisection.[40] Naturally, groups of scientists opposed her, and a general debate ensued. Much of the debate occurred under the auspices of the Royal Commission of Enquiry, which called witnesses from both sides. At this point the debate between scientists and their detractors concerning the use of animals in research was about to coalesce into a form it has basically maintained ever since. The scientific attitude was simply put by Dr. Emanuel Klein, who addressed the commission as follows:

KLEIN: I think with regard to an experimenter, a man who conducts special research and performs an experiment, he has no time, so to speak, for thinking what will the animal feel or suffer. His only purpose is to perform the experiment, to learn as much as possible and to do it as quickly as possible.

QUESTION: Then for your purpose you disregard entirely the question of the suffering of the animal in performing a painful experiment?

KLEIN: I do.

Although we need to put this in the context of the time, these words are precisely the kind of thing a practicing scientist might say today. Eventually a new amendment to the Cruelty to Animals Act of 1849 was passed in 1876. Unfortunately, although this bill paid lip service to animal welfare and protection, the scientific community had succeeded in getting it watered down to the point where nothing really changed. The bill ensured that restrictions on performing vivisection could, more often than not, be worked around by the issuing of special permissions to scientists by the Home Office.

The burgeoning movement for confronting cruelty to animals in all its forms certainly began to attract a glittering list of supporters from all walks of life, and a coalition between animal welfare groups and political entities such as Fabians and suffragettes began to develop. People like George Bernard Shaw, Mark Twain, Thomas Hardy, and Annie Besant, Madame Blavatsky's successor as head of the Theosophical Society, lent their voices to the debate, and the great illustrator Walter Crane designed the cover for the *Journal of the National Antivivisection Society*—a knight in shining armor protecting defenseless animals from the knife of a scientist (Figure 7.7).[41]

However, nothing illustrated the widespread nature of the debate like the Brown Dog affair of 1903, which exhibited all the elements of the grassroots reaction to animal use by science that we still find today.[42]

And how does such a modern confrontation look? First of all, we need some scientists who can be cast in the role of villains. In this case Ernest Starling and William Bayliss were the investigators in question. Both were experimental physiologists in the Claude Bernard mold investigating the physiological mechanisms of digestion. For this purpose, they would generally vivisect dogs. They were licensed by the Home Office, here representing the

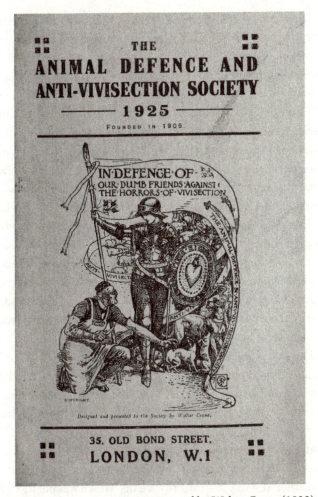

Figure 7.7   Cover of anti-vivisectionist journal by Walter Crane (1923).
Image used with permission from North Carolina State University Library, Special Collections.

evil forces of government, and used anesthesia in their investigations in what was deemed to be the appropriate manner. Both of them were also concerned with performing demonstrations of their work for medical students who were learning physiology. It should be clear that both Bayliss and Starling were first-class scientists. They discovered the hormone secretin and coined the word "hormone" to describe chemical messengers of this type. They were certainly potential Nobel Prize winners—their scientific work was of a very high caliber. Nevertheless, looked at in another way, their experiments were horribly cruel, and whereas Harvey never had to justify his experiments to the public in general, by the early years of the twentieth century there were plenty of people including an active antivivisection society who were ready to provide dissenting opinions to what the two scientists were doing. And that is exactly what happened. Two young Swedish women, Lizzy Lind af Hageby and Leisa Schartau, who were domiciled in England, infiltrated Bayliss and Starling's classes in physiology, which included numerous live demonstrations of canine vivisection. Here again is another extremely modern tactic—animal rights guerillas infiltrating scientific laboratories to free animals or wreak havoc. In one instance the two young women observed the preparation of a small brown dog for "survival surgery"—that is, the dog was vivisected, then allowed to recover, and then vivisected again, something that was against the law. Horrified by what they had seen, they published a book entitled *The Shambles of Science: Extracts from the Diary of Two Students of Physiology* (1903). A copy of the book was given to Stephen Coleridge (a descendent of the poet), a lawyer associated with the antivivisectionists. Coleridge used the material to give a well-publicized lecture in which he took the two scientists to task: "If this is not torture, let Mr. Bayliss and his friends tell us in Heaven's name what torture is?" Bayliss sued Coleridge for libel and won the case, even though it was pointed out that by preparing the dog for survival surgery he had in fact broken the law. Nevertheless, public opinion clearly backed Coleridge, and Bayliss, whatever his merits as a scientist, was cast as the villain.

Lind af Hageby founded her own society in 1906 called the Animal Defense and Antivivisection Society and collected money to fund the erection of a memorial statue to "The Brown Dog," which was put up in Battersea Park (Figure 7.8). The statue had the following inscription: "In memory of the brown Terrier dog done to death in the laboratories of University College in February 1903 after having endured vivisection extending over more than two months and having been handed over from one vivisector to another till death came to his release. Also in memory of the 232 dogs vivisected in the same place during the year 1902. Men and Women of England: How long shall these things be?"

The statue, then, was an actual physical symbol of the debate that had set scientists and members of the public at each other's throats. Trouble ensued; students acting as the shock troops became involved. Starting with small groups but eventually swelling to demonstrations in the thousands, groups of medical and veterinary students together with provivisectionists

Figure 7.8 (Left) Original Brown Dog statue of 1906. (Right) Replacement from 1985 by Nicola Hicks.
Images courtesy of Alamy and Wikimedia/with permission from Tagishsimon.

marched down the Strand and other London streets, disrupted meetings of "feminists," and demanded that the statue be removed, something that the Battersea council was loath to do. Indeed, hundreds of police were detailed to protect the statue from the iconoclasts, at considerable public expense. The trouble dragged on as the British people took sides in the Brown Dog debate, which basically pitted scientists, medics, and their supporters in the government against others from many different walks of life, although most of these were clearly from the upper and middle classes; working-class people would not be drawn into the debate as prominent antivivisectionists until after the Second World War. Then suddenly, one day in 1910, the statue disappeared. It was rumored that the Battersea council had just gotten tired of paying to defend it. It seems that it was melted down; at any rate, it was never seen again. But of course, by that point the Brown Dog had become a martyr to the antivivisectionist cause, and the disappearance of the statue produced an enormous amount of publicity in the popular press, a wave of antivivisectionist sympathies, and huge public demonstrations attended by thousands of Brown Dog supporters.

The affair eventually fizzled out but was not forgotten. In 1985, seventy-five years after the original riots, the National Anti-Vivisection Society and the

British Union for the Abolition of Vivisection, two current manifestations of British antivivisectionist sentiment, commissioned a new statue to be erected in Battersea Park along with the following inscription:

This monument replaces the original memorial of the Brown Dog erected by public subscriptions in Latchmere Recreation Ground, Battersea in 1906. The sufferings of the Brown Dog at the hands of vivisectors generated much protest and mass demonstrations. It represented the revulsion of the people of London to vivisection and animal experimentation. This new monument is dedicated to the continuing struggle to end these practices. After much controversy the former monument was removed in the early hours of 10 March 1910. This was the result of a decision taken by the then Battersea Metropolitan Borough Council, the previous Council having supported the erection of the memorial. Animal experimentation is one of the greatest moral issues of our time and should have no place in a civilized society. In 1903 19,084 animals suffered and died in British Laboratories. During 1984 3,497,335 experiments were performed on live animals in Great Britain. Today, animals are burned, blinded, irradiated, poisoned and subjected to countless other horrifying cruel experiments in Great Britain.

This naturally raised the ire of the British medical research establishment, which published a response in the *British Medical Journal*. Then, in a wonderful example of history repeating itself, the new statue disappeared in 1992; it was removed by the local council, eliciting an equally angry response by the antivivisectionists. Eventually the statue was reinstated in 1994 but in a "more secluded" spot. And there it remains to this day—unless? Even so, the current statue has been criticized by more radical elements of the antivivisectionists, who have argued that it looks too much like somebody's pet dog and won't make people uncomfortable enough. Presumably we have not heard the last of the little Brown Dog.

From our current perspective the Brown Dog affair is extremely interesting because all the different elements of the modern debate between the animal welfare community and scientists are clearly present. Naturally, the debate had something to do with science, but it wasn't a scientific debate. By the start of the twentieth century science had become a political issue that concerned all of society, not just scientists. Scientists could no longer hide away in their laboratories doing whatever they wanted to do, far from the public eye. Scientists were now accountable to public opinion and were shocked to find that it was not generally in favor of their methods. Scientists had the attitude, and still do today, that they are the guardians of the "truth" and that everybody else should listen to what they have to say because they are objectively correct. It is their opinions that are "valid" because they are the result of the use of the "scientific method." Members of the public oppose them because they are uneducated about science and don't understand what a great job it is doing. During the Brown Dog affair scientists were surprised by the fact that members of the public felt so strongly about these matters that they infiltrated their sanctums and took them to task, something that animal liberationists are still wont to do. What has changed is that the array of forces on both sides of the argument has greatly expanded in number. Nowadays it is not just a question of a few dogs and cats being dissected at medical schools opposed by a group of aristocrats and intellectuals. Millions upon millions of animals are vivisected every year on the altar of scientific truth, and millions upon millions of animal lovers are now united to oppose them.

Following the Brown Dog affair, the carnage resulting from two world wars predictably distracted people from questions about animal welfare as they focused on matters of human self-preservation. The antivivisection movement went into something of a decline. But as the world recovered, by the 1960s new philosophies of universal love of humanity and Nature were again coming to the fore, and new voices were raised to take up the banner of antivivisection once again, now producing sophisticated arguments that were appropriate for the modern age. And so now the armies of animal lovers are once more on the move and confronting science anew as the sides engage each other once again. But we need science; so how will this play out in the twenty-first century?

Notes

1. "Choupette," *Wikipedia,* July 2022, https://en.wikipedia.org/wiki/Choupette; Dana Thomas, "What Happened to Choupette?," *New York Times*, January 2020, www.nyti mes.com/2020/01/21/style/karl-lagerfeld-choupette.html.
2. *Monthly Religious Magazine*, vols. 29–30 (Leonard C. Bowles, 1863), https:// books.google.com/books?id=PXYUAAAAYAAJ; Charles Nevin, "Minor British Institutions: The Cat Flap," *Independent*, February 2011, https://www.independent. co.uk/news/uk/this-britain/minor-british-institutions-the-cat-flap-2223662.html.

3. Pet Age Staff, "APPA Report Details Pet Ownership among Generations," Pet Age, June 2020, https://www.petage.com/appa-report-details-pet-ownership-among-generations/; "Pet Ownership, Spending Going Strong," American Veterinary Medical Association, May 2019, https://www.avma.org/javma-news/2019-06-01/pet-ownership-spending-going-strong.

4. Pet Age Staff, "APPA Report Details Pet Ownership among Generations"; "Pet Ownership, Spending Going Strong."

5. Anonymous, "Four Legs Better?," The Economist, June 2019, 2244609415.

6. E. Friedmann et al., "Animal Companions and One-Year Survival of Patients after Discharge from a Coronary Care Unit," Public Health Reports (Washington, D.C.: 1974) 95, no. 4 (August 1980): 307–312.

7. Erika Friedmann et al., "Social Interaction and Blood Pressure: Influence of Animal Companions," Journal of Nervous and Mental Disease 171, no. 8 (August 1983): 461–465, doi:10.1097/00005053-198308000-00002; Suzanne C. Miller et al., "An Examination of Changes in Oxytocin Levels in Men and Women before and after Interaction with a Bonded Dog," Anthrozoös 22, no. 1 (March 2009): 31–42, doi:10.2752/175303708X390455.

8. Mark Strauss, "Americans Are Divided over the Use of Animals in Scientific Research," Pew Research Center, August 2018, https://www.pewresearch.org/fact-tank/2018/08/16/americans-are-divided-over-the-use-of-animals-in-scientific-research/; Ike Swetlitz, "Americans' Opposition to Animal Testing at Record High, Survey Finds," Statnews, May 2017, https://www.statnews.com/2017/05/12/americans-oppose-animal-testing/.

9. In ancient Greece many owners slept with their pets and there was a tradition of letting hunting dogs sleep with slaves to provide warmth.

   Louise Calder, Cruelty and Sentimentality: Greek Attitudes to Animals, 600–300 BC (Oxford: Archaeopress: Available from Hadrian Books, 2011); Liliane Bodson, "Attitudes Toward Animals in Greco-Roman Antiquity," International Journal for the Study of Animal Problems 4 (December 1983): 312–320.

10. Simon J. M. Davis and François R. Valla, "Evidence for Domestication of the Dog 12,000 Years Ago in the Natufian of Israel," Nature 276, no. 5688 (December 1978): 608–610, doi:10.1038/276608a0.

11. Bridget Alex, "Ancient Pets Got Proper Burials," Discover, September 2018, https://www.discovermagazine.com/planet-earth/ancient-pets-got-proper-burials.

12. "Alexander the Great's Dog, Peritas, Changed History by Biting an Elephant," The Blissful Dog (blog), n.d., https://theblissfuldog.com/blogs/news/91483011-alexander-the-greats-dog-changed-history-by-biting-an-elephant.

13. "Alexander the Great's Dog, Peritas, Changed History by Biting an Elephant"; Summer Trentin and Debby Sneed, "Alexander and Bucephalus," Essay, June 2018, https://www.colorado.edu/classics/2018/06/19/alexander-and-bucephalus.

14. Bodson, "Attitudes Toward Animals in Greco-Roman Antiquity"; Calder, Cruelty and Sentimentality.

15. Francis Lazenby, "Greek and Roman Household Pets," *Classical Journal* 44, no. 5 (February 1949): 299–307.

16. Stephen T. Newmyer, "Speaking of Beasts: The Stoics and Plutarch on Animal Reason and the Modern Case against Animals," *Quaderni Urbinati Di Cultura Classica* 63, no. 3 (1999): 99, doi:10.2307/20546612.

17. Joyce E. Salisbury, *The Beast Within: Animals in the Middle Ages*, 3rd ed. (Milton Park, Abingdon, Oxon; New York: Routledge, 2022).

18. Keith Thomas, *Man and the Natural World: Changing Attitudes in England, 1500–1800* (New York: Oxford University Press, 1996).

19. Salisbury, *The Beast Within*.

20. Sara Asu Schroer, "Jakob von Uexküll: The Concept of *Umwelt* and Its Potentials for an Anthropology Beyond the Human," *Ethnos* 86, no. 1 (January 2021): 132–152, doi:10.1080/00141844.2019.1606841.

21. Michel de Montaigne and M. A. Screech, *The Complete Essays*, Penguin Classics (London; New York: Penguin Books, 1993); Adam Gopnik, "Montaigne on Trial," *New Yorker*, January 2017, https://www.newyorker.com/magazine/2017/01/16/montaigne-on-trial; John Hunt and Sean Collins, "They Understand What We Say," *Joyce Project* (blog), 2017, https://www.joyceproject.com/notes/040071theyunderst and.htm; Hassan Melehy, "Montaigne and Ethics: The Case of Animals," *L'Esprit Créateur*, Montaigne and the Question of Ethics, 46, no. 1 (Spring 2006): 96–107; Philippe Desan, ed., *The Oxford Handbook of Montaigne*, Oxford Handbooks (New York: Oxford University Press, 2016).

22. Gopnik, "What Do We Really Know about the Philosopher Who Invented Liberalism?"; Hunt and Collins, "They Understand What We Say"; Melehy, "Montaigne and Ethics"; Desan, *The Oxford Handbook of Montaigne*.

23. Hassan Melehy, "Silencing the Animals: Montaigne, Descartes, and the Hyperbole of Reason," *Symploke* 13, no. 1 (2005): 263–282, doi:10.1353/sym.2006.0026.

24. "Modern History Sourcebook: John Locke (1632–1704): Some Thoughts Concerning Education, 1692," Internet Modern History Sourcebook, n.d., https://sourcebooks.fordham.edu/mod/1692locke-education.asp.

    "One thing I have frequently observed in children, that when they have got possession of any poor creature, they are apt to use it ill; they often torment, and treat very roughly young birds, butterflies and such other poor animals, which fall into their hands, and that with a seeming kind of pleasure. This I think should be watched in them, and if they incline to any such cruelty; they should be taught contrary usage. For the custom of tormenting and killing of beasts will, by degrees, harden their minds even towards men; and they who delight in the suffering and destruction of inferior creatures, will not be apt to be very compassionate or benigne to those of their own kind."

25. David Hume and Michael P Levine, *A Treatise of Human Nature* (New York: Barnes & Noble, 2005), sec. 16, https://archive.org/details/treatiseofhumann0000hume_l3g0.

26. Katie Whitaker, *Mad Madge: The Extraordinary Life of Margaret Cavendish, Duchess of Newcastle, the First Woman to Live by Her Pen* (New York: Basic Books, 2002).

27. Kathryn Shevelow, *For the Love of Animals: The Rise of the Animal Protection Movement* (New York: Henry Holt and Co, 2008), pt. 3: Speaking for Animals; Stephen Eisenman, *The Cry of Nature: Art and the Making of Animal Rights* (London: Reaktion Books, 2013).

28. Galvani's (see Chapter 3) nephew Aldini had put on public lectures demonstrating the existence of animal electricity in London at the beginning of the nineteenth century in which he "reanimated" the bodies of recently executed criminals by passing electric currents through their skulls. Some of the cadavers opened their mouths and elicited sounds. Ladies in the audience swooned. The demonstrations were widely reported in the newspapers and were probably read by, among others, Mary Shelley.

29. "Animal Rights: A History of Jean-Jacques Rousseau," *Think Differently about Sheep* (blog), n.d., http://thinkdifferentlyaboutsheep.weebly.com/animal-rights-a-history-jean-jacques-rousseau.html.

30. An Enlightenment-era word for a member of the Royal Society or similar individual who took part in scientific experimentation, as in Thomas Shadwell's 1676 satirical play about scientists, *The Virtuoso*.

31. "The Tatler: By the Right Honourable Joseph Addison, Esq," https://quod.lib.umich.edu/e/ecco/004786805.0001.000/1:53?rgn=div1;view=.

32. Andreas-Holger Maehle, "Literary Responses to Animal Experimentation in Seventeenth- and Eighteenth-Century Britain," *Medical History* 34, no. 1 (January 1990): 27–51, doi:10.1017/S0025727300050250.

33. Maehle, "Literary Responses to Animal Experimentation in Seventeenth- and Eighteenth-Century Britain."

34. Maehle, "Literary Responses to Animal Experimentation in Seventeenth- and Eighteenth-Century Britain."

35. Samuel Johnson, "No. 17. Expedients of Idlers," *Samuel Johnson's Essays* (blog), August 1758, https://www.johnsonessays.com/the-idler/expedients-of-idlers/.

36. Ariane Dröscher, "Images of Cell Trees, Cell Lines, and Cell Fates: The Legacy of Ernst Haeckel and August Weismann in Stem Cell Research," *History and Philosophy of the Life Sciences* 36, no. 2 (October 2014): 157–186, doi:10.1007/s40656-014-0028-8.

37. Shevelow, *For the Love of Animals.*

38. Shevelow, *For the Love of Animals.*

39. The first animal protection group in the United States, the American Society for the Prevention of Cruelty to Animals (ASPCA), was founded by Henry Bergh in April 1866. Bergh had been appointed by President Abraham Lincoln to a diplomatic post in Russia and had been disturbed by the mistreatment of animals he witnessed there. He consulted with the president of the RSPCA in London and returned to the United States to speak out against bullfights, cockfights, and the beating of horses. He created a "Declaration of the Rights of Animals" and in 1866 persuaded the New York state legislature to pass anticruelty legislation and to grant the ASPCA the authority to enforce it.

40. Nicolaas A. Rupke, *Vivisection in Historical Perspective*, The Wellcome Institute Series in the History of Medicine (London: Routledge, 1990); Anita Guerrini, *Experimenting with Humans and Animals: From Galen to Animal Rights*, Johns Hopkins Introductory

Studies in the History of Science (Baltimore, MD: Johns Hopkins University Press, 2003).

41. But not everybody—H. G. Wells took a provivisection position, although one wonders about that considering he wrote *The Island of Dr. Moreau*. Guerrini, *Experimenting with Humans and Animals*; Rupke, *Vivisection in Historical Perspective*.

42. Mason, *The Brown Dog Affair*; "The Brown Dog Affair" (London: Two Sevens Publishing, 1997), https://www.thehistorypress.co.uk/articles/the-brown-dog-affair/.

8

Not Just Kids

Do not unto other creatures that which is repugnant to you. Everything else is commentary.

—Rabbi Hillel, first century BCE, when asked to summarize the entire Torah while standing on one leg

The young animal sits in its cage waiting for the scientist to resume today's testing. There have been a series of these experiments. The animal failed the mirror test and showed no language communication skills or other advanced cognitive abilities. Today the scientist begins by jabbing a needle into the animal. Result: *vocalization and strong escape response to nociceptive intervention.* This is good. Conclusion: *animal is suitable for use in experiments on pain and pain-reducing medications.* So now what? The scientist considers his options based on the projects he has funded by the government. Perhaps he should break the animal's leg or back or smash its skull and then examine the properties of the resulting pain? Yes, he likes that idea. He can imagine the sort of data he will get: highly significant data points illustrated in the pastel hues favored by today's scientific journals. He mentally licks his lips in anticipation. No matter that the animal in the cage is a one-year-old human child—the scientist knows that if an animal has no meta-consciousness, no ability to communicate using language, and no apparent ability to reason, it has no moral status and can therefore be used for any research purpose whatsoever, provided it has been approved by the National Institutes of Health (NIH). It is certainly the case that the experiments will have to be clearly described and approved by the university's Institutional Animal Care and Use Committee (IACUC). But that won't be a problem because none of the proposed experiments fall outside of the law. He will be given the green light and be able proceed as planned, of that he is certain.

Critiques of these kinds of narratives are usually known as arguments from marginal cases (AMCs). It is quite clear that if the experimental subject

The Rise and Fall of Animal Experimentation. Richard J. Miller, Oxford University Press. © Richard J. Miller 2023. DOI: 10.1093/oso/9780197665756.003.0008

in the situation described was a mouse, there would not be a problem. In fact, there are lots of scientists who break the legs, backs, or skulls of mice for experimental purposes. Indeed, such things do in some respects reflect real-life situations that humans experience when they suffer traumatic injuries. Nevertheless, if the experimental subject was a young child or an advanced Alzheimer's patient, whose language and cognitive skills were just as "impoverished" as those of a mouse, they would certainly not be treated in this way. There appears to be an inconsistency with these arguments unless we decide that we won't perform experiments on animals or that we will perform experiments on cognitively impoverished humans. In other words, we seem to have reached something of a logical impasse.

Fortunately, this scenario is fiction. However, it's the kind of "what if" situation that philosophers like to consider when they are thinking about animal experimentation. After all, these are the kinds of arguments that scientists have used for centuries when trying to justify animal research: that animals are inferior to humans in key respects, that animals are property, and that it is therefore reasonable for us to use them in our studies for any purpose we think is appropriate. But clearly, in this instance, one can see that the same thing would have to apply to a very young human or perhaps a very old person who is suffering from advanced dementia—and no, we wouldn't perform experiments like these on humans, whatever their cognitive status. We feel that we have a moral obligation toward humans that we don't have toward mice. At the very least, these considerations should give us pause for thought.

And they do give us pause for thought nowadays—lots of it. Prior to the Second World War, the ethical appropriateness of animal experimentation was something that was only considered to be of importance by grassroots organizations—groups like the Brown Dog antivivisectionists ; it wasn't a subject of academic interest. This couldn't be further from the truth these days. Like race and gender, the fate of abused animals is something that many people think about nowadays, and this is reflected by the way academic institutions view the topic. Many universities now have entire departments devoted to animal studies that examine different aspects of the interactions between humans and animals and cover things like animal experimentation and also factory farming, conservation, animal sports, zoos, and a wide variety of other subjects. Moreover, many highly respected academic presses publish book series devoted to animal-related issues, and numerous universities teach courses on animals and the law. There has been a revolution in our thinking about animals, particularly since the 1970s, so that animal studies is now a highly respected discipline for academic discussion. One result of all of this is that, since the Second World War, a great deal has been written by philosophers, ethicists, and legal scholars on subjects like animal experimentation. Unfortunately, these studies are generally restricted to the humanities, which are frequently the purview of campuses that are separated from medical schools and the university departments that study biomedical sciences and are actually responsible for carrying out animal experimentation. And, of course, the opposite is also true. Academics in the humanities are often not up to date with the latest scientific advances and, perhaps more problematically, what it is like to live the everyday life of a research scientist and what is behind the day-to-day decisions that scientists make. There is the issue of not only physical distance but also intellectual distance. The majority of scientists are not exposed to issues concerning the ethics of animal research; it simply isn't part of their education. If anything, scientists are actively discouraged from engaging with these topics. For several years, I taught a course to graduate students in the medical school at my university on the history and philosophy of science, which covered topics like ethical choices in science. The course was popular with the students, and a lot of them signed up for it. Eventually, however, it was "suggested" that I stop teaching the course. The reason was that a lot of PhD mentors complained that their students were taking a course that wasn't teaching them about "real science" and that they should do something more useful with their time. What is more, many scientists have also come to regard these sorts of courses as threatening. As the philosopher Bernard Rollin has written:

> Historically, the scientific community—at least in the USA—did not perceive the use of animals in research as an ethical issue. Anyone who raised questions about the way animals were kept and treated during experiments ran the risk of being stigmatized as an anti-vivisectionist; a misanthrope preferring animals to people; or an ingrate who did not value the contributions of biomedical science to human health and well-being.[1]

Medical research is a vital revenue stream for most universities, and anything that might interfere with it is regarded as anathema to university administrations. Consequently, however finely honed philosophical arguments about the ethics of animal research may be, they make practically no difference to the scientists who carry out experiments on animals as part of their daily routine. But shouldn't scientists be interested in what philosophers have to say? Perhaps they are missing something important? All this scientific progress we hear about—what is it really? How do we view it and evaluate it? As I have argued, animal experimentation is rapidly becoming redundant, and in the not-too-distant future, the subject will be moot because I don't think good scientists will be performing experiments on animals anymore. But that is not the state of affairs at the moment, and the subject is an important one for helping us frame our attitudes toward animals and the world in general. When we think about these issues, there are probably two major areas that should interest us—the laws that concern animal welfare and what factors should determine the code of ethics that governs the use of animals in research. As has been said, "Laws tell you what you are allowed to do, and ethics tell you what you should do."[2]

As we have seen, there have certainly been times throughout history when animal research has been responsible for some of the most important scientific advances ever made. Harvey's discovery of the circulation of the blood and Bernard's studies on homeostasis made extensive use of vivisection. There can be no doubt as to the scientific importance of experiments like these, even though at the same time, we recognize that what these scientists did was extraordinarily cruel. But nobody seriously questioned whether Harvey or Bernard should be allowed to do the things they did. Experiments like theirs, at least carried out exactly as they were originally performed, would not be permitted these days. Starting in the nineteenth century and progressing into the twenty-first century, there have been some important changes in the laws that regulate how scientists conduct animal experiments (see Chapter 7). These changes were driven by public outrage over revelations concerning the treatment of animals in research laboratories. Moreover, academic philosophers and ethicists have weighed in on these matters and have suggested codes of behavior that they think are important for ensuring the ethical treatment of animals. The entire subject is one that is now hotly debated (except, unfortunately, by scientists), and numerous different organizations representing a wide spectrum of opinions are struggling to have their voices heard and to influence the situation.

As we have seen throughout history, there have been writers who had different opinions about the treatment of animals. In the ancient world, we may read about such views in the writings of Aristotle, the Stoics, and Galen on the one hand and Pythagoras, Plutarch, and Porphyry on the other. In the Middle Ages, St. Augustine and the Thomists were extremely influential. In the seventeenth century, when science began to spread its wings once again, we see the opinions of Descartes on the one hand and Montaigne on the other. And so, we arrive at the period just prior to the development of recognizably modern science in the nineteenth and twentieth centuries, when some of the basic assumptions that historically governed our attitudes toward the interactions between humans and animals began to come into question. A particular problem with most of the arguments made in previous centuries, beginning as far back as Aristotle, was that they all started with the basic assumption that humans were "better" than animals, that animals were simply not worthy of the same kind of moral consideration that would be accorded to human beings. These views, as we have seen, were based on a collection of ideas that attribute uniqueness to human beings, setting them aside from the rest of the animal world. Only human beings were made in the image of the Almighty and could go to heaven, only human beings were rational and conscious, and so on. This kind of approach is known as human exceptionalism, anthropocentrism, or speciesism.

If we begin with the idea that the biological uniqueness of humans also implies that they should be granted exceptional moral consideration, then animals have pretty much had it. However, these kinds of arguments eventually run up against a brick wall. Let us consider the views of the eighteenth-century German philosopher Immanuel Kant, one of the most influential philosophers of all time. Kant certainly thought and wrote about the moral status of animals and what that implied for the way we interact with them and with each other. According to Kant, "So far as animals are concerned, we have no direct duties. Animals are not self-conscious and are there merely as a means to an end. That end is man."

Kant concluded that animals were not "independent autonomous agents" and therefore had no moral standing. Hence, animals could not be wronged by humans. Kant wrote:

> If a man shoots his dog because the animal is no longer capable of service, he does not fail in his duty to the dog, for the dog cannot judge, but his act is inhuman and damages in himself that humanity which it is his duty to show towards mankind. If he is not to stifle his human feelings, he must practice kindness towards animals, for he who is cruel to animals becomes hard also in his dealings with men.

A dog, then, according to Kant, is not the sort of thing that can be wronged by a human, even in principle. Kant's views are somewhat similar to those of the Thomists, that there is nothing wrong with the mistreatment of animals per se, but we shouldn't behave this way because it would ultimately influence human behavior and how we treat one another. This is certainly a valid point. It is true that humanity is degraded through the mistreatment of animals, but that isn't the main reason we should treat them with kindness and respect.

What does Kant mean when he says that animals are not "independent autonomous agents"? The concept of autonomy embraces a variety of different ideas suggesting that an entity can act independently in a goal-directed manner resulting from the workings of a rational mind that is self-conscious and capable of assessing the results of its own actions. These are certainly characteristics associated with humans but not, according to Kant, associated with animals.

I think the conclusions that Kant reached would be rejected by most people today as being counterintuitive, to say the least, because they contradict all the current scientific evidence suggesting the sentience and consciousness of animals, which, of course, was not yet available in Kant's time. It is clear that many animals, at least animals that have even a rudimentary nervous system, are sentient in some fashion and, unbeknownst to Kant, have complex cognitive abilities reflected by sophisticated communication skills, even if these aren't human language, as well as rational abilities, even if these are animal specific. As I argued, animals need to be appreciated *on their own terms* and for what is good or bad for them in an animal-appropriate manner and not what would be good or bad for them if they were viewed as types of inadequate human beings. Needless to say, kicking or shooting a dog is bad for the dog in its own way, of that there is very little doubt, and we don't have to justify this fact based on comparisons with humans. Clearly, the results of scientific investigations on the cognitive capabilities of animals that have occurred subsequent to Kant's lifetime mean that more modern ethical frameworks are required these days when thinking about animals that reflect twenty-first-century scientific knowledge and sensibilities. So, Kant is an interesting figure if we want to study the history of philosophy but not very relevant for dealing with animals in today's world.

Writing and thinking about the moral status of animals began to pick up after the Second World War. Nazi philosophy had stripped Jews of any kind of moral status, reducing them to the equivalent of laboratory animals allowing Nazi "scientists" to perform any kind of experiment on them with no ethical constraints whatsoever, something they did with truly ghastly consequences. And there were other revelations as well. It turned out that even countries like the United States were not completely free from skeletons in their closets. Revelations about the Tuskegee syphilis studies, the CIA MK-ULTRA brainwashing experiments, and a number of other clandestine operations prompted laws to be passed regulating exactly how and when humans could participate as subjects in scientific investigations. New laws required that an institutional review board (IRB) be established by every research institution to safeguard the concept of "informed consent"; human subjects could only participate in experiments if they were adequately informed as to what was going to happen and gave their explicit permission of their own free will. Additionally, by the middle of the 1970s, discussions about the moral status of animals started to become part of a renewed appreciation of the natural world in general. The zeitgeist of the 1960s began to favor the view that we should be *in loco parentis* when it comes to our planet and the creatures that inhabit it. Influential texts included the publication of Rachel Carson's *Silent Spring* in the 1960s, Jane Goodall's books describing her observations on the lives of apes in the wild in the early 1970s, and the development of the Gaia hypothesis by Lovelock and Margolis, which cast the entire planet Earth as a vast living organism. The publication of *Animal Liberation* by the philosopher Peter Singer in 1976 threw down the gauntlet to those who persisted in cruel and unacceptable treatment of animals for no good reason.[3] I say "for no good reason" because Singer was not completely against the use of animals in areas such as experimental research if, in fact, there was "a good reason." Singer is the intellectual heir to the nineteenth-century utilitarian philosophers such as Jeremy Bentham and John Stuart Mill, who generally judged the moral worth of actions according to the dictum "the greatest good for the greatest number." Hence, they thought that justification of our actions should be determined by examining all their consequences, assigning values to them, and deciding if the ultimate balance was positive or negative. As we have seen, Bentham extended this idea to animals as well as humans, particularly with respect to the areas of pleasure and pain, the things that sentient creatures strive to maximize or avoid. Bentham clearly thought that in most cases it would be difficult to show that the mistreatment of animals would lead to some general good, but he didn't rule it out.

If we think about it, sometimes it seems that animal research might be justified. Consider our current coronavirus epidemic as an example. We have needed to develop a vaccine "at warp speed" in order to fight an infection that has done a great deal of damage and created suffering for millions of people throughout the world. This seems like a good test case for deciding whether animal studies can sometimes be justified. Surely, it might be argued, if we need to use monkeys or mice to develop a vaccine that could save the lives of millions of humans, such a thing would be justified if anything ever was. How could even the most ardent animal welfare advocate argue against such a conclusion? Surely the utilitarian balance must fall on the side of approving animal research in this instance. We have been down this road before: in the panic that surrounded polio in the 1950s, literally millions of monkeys were used to develop a vaccine that helped defend humans against a horrible disease. In fact, so many animals were used that it caused the population of monkeys in India to drop from hundreds of millions to hundreds of thousands.[4] There was very little outcry against doing monkey research at the time, so if we are going to deny the use of animals for COVID-19 research, we better have a very good reason. As things turned out, any utilitarian arguments favoring animal research in this case were trumped by practical scientific considerations; animal studies just didn't turn out to be very useful in creating a vaccine against SARS-CoV-2. At the time of writing (November 2021), a review article in the journal *Nature* described the different animal models of COVID-19 that were in play, including mice, rats, hamsters, ferrets, monkeys, mink, cats, dogs, bats, chickens, and ducks.[5] Unfortunately, none of these models precisely mimics the human disease in many key respects. Indeed, the authors wrote, "However, over the past decade, advances in engineering, cell biology and microfabrication have come together to enable the development of new human cell-based alternatives to animal models. In this regard, micro-engineered organs-on-chips and lung organoids have been shown to support key hallmarks of the cytopathology and inflammatory responses observed in human airways after infection with SARS-CoV-2 and have served to facilitate the study of human disease pathogenesis, test candidate COVID-19 therapeutic agents and expedite drug repurposing." This is precisely the view I put forward when I talked about the advances in the use of human-based organoid and other experimental paradigms for conducting research that is supposed to benefit humans. Although a very limited number of the preclinical "proof of concept" studies on the development of SARS-CoV-2 vaccines were performed in animals such as mice and monkeys, the United States and other governments encouraged vaccine makers to move straight from initial vaccine development to humans without going through all the usual animal safety trials that would usually have been required. These normally take a considerable amount of time and require the use of huge numbers of animals. Moderna's chief medical officer Tal Zaks commented about their groundbreaking vaccine: "I don't think proving this in an animal model is on the critical path to getting this to a clinical trial." This was a clear recognition of the fact that, in the end, if we want something that works in humans, examining its effects on the human population is really the only reliable test that matters. And I think that these are the kinds of arguments that will, more often than not, direct the way biomedical research is done in the future.

The Value of Scientific Publications

The problem with all the utilitarian approaches to animal experimentation is that they force us to make some kind of quantitative estimate as to the value of different experiments and their outcomes in terms of their relative consequences for animals and humans. This is the type of calculation that scientists are supposed to make to justify their research efforts. The argument that is made is usually something along the lines of, "Well, we don't know which experimental results are going to be important in the future, so how can we judge their value?" A scientist will inevitably say, "You don't know whether my recent paper in the AJTR (*Antarctic Journal of Toenail Research*: impact factor minus 12), which used 500 mice, isn't going to be vital in our fight against . . . [fill in the name of some important disease from the following list—cancer, Alzheimer's disease, stroke, arthritis, diabetes, etc.]." In other words, the research has some indeterminant potential value that tips the utilitarian balance in favor of the animal experiments carried out in this particular instance. However, it should be pointed out that we are comparing a certainty with a possibility. The 500 mice are dead, that is certain, but the paper in the AJTR is just potential and may ultimately prove to have no value whatsoever.

The truth of the matter is that these days, the value of most scientific publications really has little to do with science and much more to do with career building by the scientists who publish them. Most scientific publications have a status that, in many respects, is reminiscent of the "readymades" created by the artist Marcel Duchamp. Back in 1917, Duchamp presented a work called "Fountain" to the Society of Independent Artists for display at their upcoming exhibition. In the event, "Fountain" turned out to be a

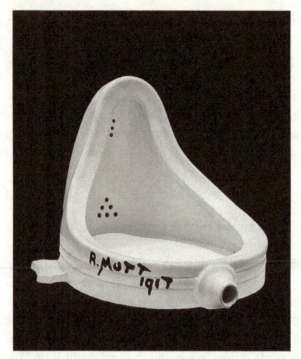

Figure 8.1 Duchamp's "Fountain" (1917).
Image from Alamy.

mass-produced common or garden urinal stuck on a pedestal (Figure 8.1).
Duchamp's "Dadaesque" gesture raised many questions about the nature of
art, which are still discussed today. However, there are certain points about
what he did that are relevant to the current discussion.

In my opinion, many scientific papers share with "Fountain" the fact that
they have little to show in terms of technical virtuosity and have the aura of
mass production. The main thing that validated "Fountain" as art was that
it was accepted by the organizers of an art exhibition; so, if it's in an art exhibition, it must be art—correct? This is also true of the majority of scientific publications. They are termed "science" because they are published in
scientific journals, which legitimize them as science. If they are published
in scientific journals, they must be science—correct? Unfortunately, this really isn't the case. The life of a scientist is very much a "publish or perish"
affair, and scientists who want to get promoted, receive tenure, garner research grants, and make their way in the world need to publish papers,
come what may. There is an old joke in medical research that goes like

this: Question: "How do you define a drug?" Answer: "A drug is a substance that when injected into a mouse generates a publication." One response to this need is that the world of scientific publishing has expanded enormously, creating a huge number of journals where scientists can publish their results no matter how good or bad they are, and this does garner them a sort of respect when their efforts are added up by tenure committees and other bodies. How many times have I heard in the media that a particular claim is backed up by science because a paper was published in journal X authored by scientist Y? Sadly, statistics suggest that many scientific publications are never read by anybody except the scientists who publish them and are rarely cited in other scientific work except by the authors themselves.[6] As far as the fate of the animals used in scientific investigations is concerned, a recent survey determined that the results of experiments on animals at the University of Utrecht, a very typical research university, were hardly ever reported in literature. Of the 5,500 mice that were requested to be used in the experiments covered by the investigation, results on only 23% were ever reported in the scientific publications, indicating a complete waste of money, effort, and animal lives.[7]

I suspect that many publications have little value from the point of view of their contribution to scientific knowledge, even if they may have some value in terms of promoting scientists in their careers, or what we might call "having a life in science." In other words, many papers published today probably really only have value because they are commodities rather than because they generate any important new knowledge. In truth, apart from in very exceptional instances, the ammunition that scientists use to justify their animal research from the utilitarian point of view is just a cartridge full of blanks. The correct view of today's scientist is someone who is busy trying to parlay research data into funding rather than some erudite person trying to investigate the secrets of the universe. Nevertheless, utilitarian arguments do leave the door open; some research that uses animals may be of value and may be useful. Still, even if we can make a utilitarian argument for the value of certain animal research, does this really mean we should do it? Don't forget, Herophilus and Erasistratus (see Chapter 2) found out a great deal about cardiovascular physiology simply by killing some "criminals" who were going to be executed anyway. Was that okay? Very few people nowadays would agree that it was. In other words, just because some research does provide important results doesn't mean it is ethically justified. Shouldn't the same standards apply to both humans and animals?

These problems have given rise to the other main philosophical argument about whether we should use animals in research, developed by the ethicist Tom Regan.[8] This approach doesn't require us to make quantitative value judgments concerning research and its outcomes. Regan's consideration of animals begins with the possibility that they have a significant mental life, and, as we have discussed, this applies to many kinds of animals, since most animals with a nervous system will have some degree of sentience. According to Regan, such an animal has a "life" of which it is aware. This life means something to the animal. We can see these things in action by observing the animal's intentions and actions that serve to maintain its life. If an animal is the "subject of a life," then it has a "right" to maintain its life, and attempting to deprive it of its life and/or interfering with it in a negative manner is therefore interfering with its rights. Such arguments have led to several high-profile legal cases in the United States in which it has been argued that, like humans, animals should be regarded as "persons," and, if they are persons, then there are legal consequences that result from this designation, such as being a candidate for a writ of habeas corpus so that they may be freed from captivity. These possibilities have been widely discussed, for example, in the case of Happy the elephant, who lives at the Bronx Zoo in New York. Happy is clearly very intelligent and did a stellar job passing the mirror test (see Chapter 5). Nevertheless, her suit was dismissed and, as yet, arguments like these have gained little traction within the US legal system.[9]

Views such as these are not as interested in trying to improve the life of a research animal who is going to be subjected to harmful procedures and ultimately put to death. Rather, the view is that all these things are wrong, and we shouldn't do them. There is no reason to try and add up the relative good or bad so that one can make a utilitarian value judgment about the appropriateness of animal experiments. If we do things that interfere with the rights of a creature that is "subject to a life," then that is wrong. This, then, is a more extreme view of how we should treat animals in research and basically makes much more sense. Wrong is wrong, and that is that. In the experimental scenario discussed at the start of this chapter, Regan would argue that we shouldn't do such experiments no matter whether the subjects are young children or mice. But of course, even though it might be easier to decide what to do based on these considerations, people still argue about things such as which animals are really the "subjects of a life." Humans—of course. Apes— clearly. Mice—some would argue that this isn't as clear. Octopuses—who knows? However, as we have discussed, it is now perfectly clear that all these intellectually sophisticated creatures are "subject to a life" and that good and bad things can happen to them, even if their internal psychological landscape is very different from that of a human.

There has been a great deal written about these matters since the 1970s. There are many authors who have suggested different variations of the utilitarian or rights-based approaches or combined models that incorporate "sliding scales" of values depending on the animal in question and the particular situation under consideration.[10] But, suffice it to say, there is no consensus about how to approach the matter of treating animals as research objects. We should note, however, that somebody like Harvey was limited by the available technology of the time when he carried out his experiments in the seventeenth century. Once he had decided to proceed with animal experiments, there weren't any alternatives. But as I discussed in Chapter 6, there are now many alternatives for performing experiments that don't use animals and that can be directly translated to humans in both the clinical and basic science arenas.

Desensitizing Science

The fact is that all of the approaches we have discussed have attempted to wrestle with the problem of how to cope with the concept of human "exceptionalism," the idea that not only are humans different in some respects from other creatures but also they are *sui generis* and in a completely unique category; that there is no seamless link between the rest of the animal world and ourselves of the type Darwin proposed; and that this gives humans a unique moral status, allowing them to play by different rules. As we have seen, humans have been obsessed with this idea since the time of the Greeks. We have always been at pains to stress our own importance in relation to every other thing on the planet, and this doesn't just mean animals but literally everything—every plant, every stone, every river. This is the context in which science has developed, but did it really have to be this way?

Science isn't just a set of methods for doing things like performing experiments; it's a culture, and the culture of science has a lot to say about how science goes about its business. There is almost always more than one way of answering any scientific question, and the path a scientist takes is usually dictated by the prevailing opinions of science as culture and not just science as experimental method. The nature of science as culture is very much connected with the nature of the research environment in which science is performed. This became rapidly apparent when research institutions, primarily research universities, developed in the nineteenth century. In his book *Organizing Enlightenment*,[11] Chad Wellmon quotes the opinions penned in 1810 by Friedrich W. Zimietzki concerning the burgeoning new research universities in Germany:

> For Zimietzky, science was not just a collection of facts and ideas. It was, as he put it, an "outlook on life." It was a particular way of life with its own "customs" and virtues that bound a community together and sustained an institution. For the "brothers" of the university, scientific knowledge was a common good." It was something beyond themselves, and its pursuit required vigor, intellectual freedom and, above all, love that drove them toward the "good and true" of science. Science was a distinct culture with its own authority and virtues. And it was the embodiment of the university.

The use of animals as research tools is not just the result of deciding which scientific methodologies constitute the best scientific experiments and deciding the answer to a set of philosophical questions, but more the result of what the culture of science has decided is the thing that everybody should be doing at a particular point in scientific history. If scientists are going to stop performing experiments on animals, then it is the culture of science that needs to change as well as the technology of science itself, and there are many issues that might determine whether, when, and how this might happen.

The reality is that when a scientist is confronted with the realization that they are about to injure or kill an animal, this isn't really a situation for engaging in something like a utilitarian calculus. Rather, it is a moment of high drama, a matter of life or death, a mystical instant charged with emotion. The scientist is like the emperor in an arena full of gladiators; is it thumbs up or thumbs down? If you are going to make a life-or-death decision about a sentient creature, then you should really be thinking about what you are doing. Not to do so would just be callous, and would we want a community of scientists who acted in such a manner? Surely, when a scientist considers the reality of an animal on which they are about to perform an experiment, they cannot avoid the conclusion that their actions are cruel; scientists are people, after all, and have basic feelings of compassion that are common to all human beings, and no amount of weighing the pluses and minuses of the overall value of an experiment to humanity at large can completely account for or rationalize decisions like these.

In practice, the culture of science usually deals with these situations through a process of desensitization. When a new graduate student enters a laboratory, there is a ceremony, which involves the student killing an animal, something they may well never have done before. Indeed, this is something that is usually officially sanctioned, and in many institutions, the student must go through the rite of killing in a supervised animal facility while being observed by "officials," ostensibly to ensure that the animal is killed in the "correct" fashion. But of course, this also allows the student to cross the Rubicon of personal animal killing, a ritual that, once performed, may make subsequent procedures of this sort carried out in the service of science easier to countenance. Moreover, there is also a realization that a career in science will not involve performing such acts forever once a period of training as a graduate and postdoctoral student is over. At such a point, once a scientist has been appointed to a faculty and has their own laboratory, all the practical work will typically be delegated to technicians or the aforementioned students. The scientist in charge of the work will have an office that is separated from the laboratory and may even be in an entirely different building. The scientist may hardly ever enter the laboratory and observe research in progress. At this point, the scientist is practicing "science at a distance," completely shielded from the flesh-and-blood realities of actual experimentation. An animal becomes just a dot on a graph, a theoretical entity devoid of any actual animality that the scientist can move from one place to another on their computer screen, rather like engaging in a videogame where "kills" can be added up in the hope of reaching "the next level." This process of disengagement of the scientist from the realities of experimentation cannot be expected to result in any semblance of moral behavior with respect to animals as experimental subjects, and a consideration of rights or values just misses the point. When a scientist is responsible for cruelty to animals during an experiment, they need to be actively engaged in this process in the moment, not just scientifically but also emotionally, and to seriously contemplate their actions and their consequences. If scientists aren't doing that, they should be. Different approaches are required so that the scientists of the future maintain a connection with their research endeavors in all respects, including the way they treat animals. It isn't really a case of "just the facts"; biomedical research isn't physics.

Eastern Attitudes

There are certainly other traditions that we can examine as exemplars of different attitudes toward animals and their potential use as experimental subjects. This is particularly true if we think about Eastern cultures, where obsessive anthropocentric thinking was never the order of the day, until recently at any rate. Now, Western ideas about science have made important inroads into the East. Contemporary China is described these days as a "scientific powerhouse," and India is a huge growth area for biotechnology, but, of course, it is Western science that people in these Asian countries are involved in performing. We have plonked science down in the East along with Coca Cola and Disney, and, generally speaking, conducting modern scientific research, which has a fantastic record of solving practical problems and giving us insights into the secrets of the universe, has been, and will continue to be, one of mankind's greatest cultural achievements, wherever it is performed. The issue here is its modus operandi, in particular, the use of animals in this process, not the thing-in-itself. The question is how to create scientists producing cutting-edge research through the use of ethical procedures and methodologies. Transferring our scientific culture to the East has only succeeded in producing scientific endeavors in these countries that mimic Western scientific practices and that come with the same ethical problems.

However, perhaps there are important things that we could learn in return. Here, the East may provide a counterbalance to the intellectual steamroller of Western science. Lest we forget, influential ideas have often moved in the opposite direction, from East to West, from India to the likes of Helena Blavatsky, Herman Hesse, Aldous Huxley, and many others. And if we look at these ideas, we may find clues that can guide us in our treatment of animals in research and our attitude to other creatures in general. Thinking that developed historically in Eastern countries like India was never obsessed with establishing why humans were uniquely important because they were capable of language or reason or were conscious. Because of this, Eastern attitudes toward animals didn't have to incorporate these principles—quite the opposite. The Buddhist principle of *ahimsa*, nonviolence or nonharm, can be summarized as saying that "no breathing, living, sentient creature should be slain, nor treated with violence, nor abused, nor tormented, nor driven away" and is applicable to humans and nonhumans alike.[12] Indian religions encourage the development of a compassionate attitude and empathic feelings toward all creatures so that an adherent of the faith feels that harm done to other humans or animals is harm done to oneself. Although the concept of ahimsa is important to some degree in all the Vedic religions, it is particularly important in Buddhism and Jainism and is regarded as one of the most important virtues that an adherent of these faiths can practice. It is the most important virtue of Buddhist moral philosophy, and it has profound practical consequences. Almost every Jain community in India runs an animal hospital to care for injured and abandoned animals. They don't eat animals, and they don't sacrifice them. Nonviolence is the most important concept in the Buddhist tradition for attaining peace, which must be achieved if an individual wishes to attain a state of nirvana, where one is freed from a life of suffering (*dhukka*) by release from the cycle of death and rebirth. In Buddhism and Jainism, the importance of nonviolence is based on the equality of all creatures. Every living being should be free to enjoy the full potential of its life, and this doesn't just apply to actual violence but also the concept of it. Violence that arises out of greed, hate, anger, and ignorance is considered to be the most important factor in causing suffering. Eastern religious attitudes toward animals do not require adherence to one of the Vedic faiths. History makes it clear that an appeal to the principles of compassion and nonviolence is likely to be attractive to many people in the West, who would certainly be prepared to apply it to the treatment of animals. Indeed, such an approach began to be seriously considered in Western countries soon after the initial translations of Indian religious texts began to appear, and this was particularly true of the influential philosopher Arthur Schopenhauer.

Arthur Schopenhauer (1788–1860) was greatly influenced by the idealistic philosophy of his compatriot Immanuel Kant and also by Buddhism.[13] Indian thought arrived in Europe via two major routes. The first was through British scholars associated with the East India Company, starting with an English translation of the Bhagavad Gita by Sir William Jones published in 1784, four years prior to Schopenhauer's birth, and this was followed by translations of Indian poetry and literature. Goethe, for one, was greatly impressed by these works. The second major influence on European scholars was a translation into Latin of the Upanishads in 1801, a text that was a great favorite of Schopenhauer's; he is said to have always kept a copy of the book and a statue of the Buddha on his desk. Schopenhauer wrote of his introduction to Indian thought in a letter in 1851 about a meeting he had with Goethe in Weimar in 1813, saying, "At the same time, the orientalist Friedrich Majer introduced me, without solicitation, to Indian antiquity, and this had an essential influence on me."

From Buddhism, Schopenhauer developed the idea that the struggle for life, which results from the "will to life," is the cause of suffering, a similar idea to the Buddhist concept of dhukka. Schopenhauer believed that all forms of life are conscious, and they all suffer because their will produces a continuous striving for more life, which can never be fully satisfied, so the struggle never ends. Schopenhauer concluded that death was ultimately the only way out of the problem.

For Schopenhauer, the fundamental difference between animals and humans resulted from differences in intellect, not will, which was associated with all forms of life. Schopenhauer believed that to unduly increase the suffering of another being was unethical. Why should anyone have the power to increase the suffering of others? For Schopenhauer, morality was not the result of reason, but rather the result of feelings of compassion for all creatures. His dictum of morality might be encapsulated by saying, "Harm no creature and help others as much as you can."

Because of the influence of Buddhist principles, Schopenhauer stands out as one of the first Western philosophers to accord not only moral standing but also moral rights to animals. He wrote:[14]

> In Europe a sense of the right of animals is gradually awakening in proportion to the fading and vanishing of the strange idea that the animal world was brought into being merely for humans' use and amusement as a result of which animals are treated just like things.
>
> . . .
>
> Thus, because Christian morality leaves animals out of account . . . , they are at once outlawed in philosophical morals; they are mere "things," mere means to any ends whatsoever. They can therefore be used for vivisection, hunting, coursing, bullfights and horse racing, and can be whipped to death as they struggle along with heavy carts of stone. Shame on such a morality that is worthy of pariahs, chandalas and mlecchas, and that fails to recognize the eternal essence that exists in every living thing, and shines forth with inscrutable significance from all eyes that see the sun.
>
> . . .
>
> Since compassion for animals is so intimately associated with goodness of character, it may be confidently asserted that whoever is cruel to animals cannot be a good man.

. . .

The assumption that animals are without rights and the illusion that our treatment of them has no moral significance is a positively outrageous example of Western crudity and barbarity. Universal compassion is the only guarantee of morality.

Here, then, is the important principle of nonviolence or ahimsa writ large. The man who is cruel to animals cannot be a good man, says Schopenhauer. The scientist who is cruel cannot be a good scientist. When we perform an experiment using an animal, we are involved in an act of violence. By doing so, we are contributing not only to the suffering of the animal but also to our own suffering and that of the world in general. If scientists were trained to reject violence, then surely they would be better scientists. What is more, adhering to a principle of nonviolence does not require a scientist to be religious, something that is anathema to most of them and doesn't necessarily come with a lot of metaphysical baggage. It is a basic human value. When we create the scientists of the future, they will need to receive training that makes them genuinely engage with the spirit of ahimsa or nonviolence when they consider their experimental animal subjects. This should not just be a "rubber stamp" kind of engagement that scientists participate in today but a true spiritual exercise that results from genuine feelings of compassion. Scientists who are trained to respect the lives of animals and the principle of nonviolence will therefore think very carefully about animal experimentation and perhaps cease to do it for the most part. This course of action will ultimately not come from the outside, from laws and rules that tell scientists what to do, but from the inside; it will be the result of their own moral understanding of the situation. It should be inevitable that scientists act in this way, that they are morally obliged to treat animals with respect.

Creating a new ethical science will entail not only changes in the way that scientists are educated and the motives and behaviors of scientists themselves but also changes in the structure of universities and the manner in which research funding is distributed by the government (see Chapter 9). Better ethical training of scientists is something that will be difficult to achieve as things stand at the moment because research scientists operate solely according to career-building motives and will only regard such behavior as an impediment to their primary goal of raising money. What today's scientists will respond to are changes in the laws that determine what they can do, and the law is something that can be influenced by the opinions of people at large. This, then, primarily requires efforts and actions that are external to the scientists themselves. The question is, then, can enough people engage in the principle of nonviolence as a political force to make a practical difference to the treatment of animals and the fate of animals in research in particular? One might imagine that in today's societies, which, like today's scientists, are very much driven by the concepts of competition, success, and self-interest, such an attitude would be difficult to "sell." Interestingly, Richard Wagner, who was in many respects a follower of Schopenhauer's, referred to this issue in a letter he wrote to his local antivivisection society upon joining it, although, being Wagner, he had to temper it with a little antisemitism.

> Yesterday I officially became a member of the local Society for the Protection of Animals. Until now I have respected the activities of such societies, but always regretted that their educational contact with the general public has rested chiefly upon a demonstration of the usefulness of animals, and the uselessness of persecuting them. Although it may be useful to speak to the unfeeling populace in this way, I none the less thought it opportune to go a stage further and appeal to their fellow feeling as a basis for ultimately ennobling Christianity. One must begin by drawing people's attention to animals and reminding them of the Brahman's great saying, Tat twam asi (That art thou),—even though it will be difficult to make it acceptable to the modern world of Old Testament Judaization. However, a start must be made here,—since the commandment to love thy neighbor is becoming more and more questionable and difficult to observe—particularly in the face of our vivisectionist friends.[15]

Recent history has demonstrated that an attitude that embraces nonviolence can catch on. Since Schopenhauer's time, the acceptance of Eastern ideas has become quite widespread in Western cultures, even to the extent of effectively supporting political endpoints. Needless to say, the nonviolence practiced by Gandhi was a considerable influence, particularly on Martin Luther King and others, and both of these leaders excelled in using the principle of nonviolence to serve their political agendas.[16] King made his view of animals perfectly clear: "One day the absurdity of the almost universal human belief in the slavery of other animals will be palpable. We shall then have discovered our souls and become worthier of sharing this planet with them." Indeed, there is a long-standing influence of Eastern ideas on

American thinking, from the transcendentalist movement in the nineteenth century, to the influence of Buddhist ideas on the likes of Alan Ginzburg and the Beats, to the powerful influence of the Maharishi Mahesh Yogi on the counterculture movement of the 1960s, when "peace, love, and flowers in your hair" was the mantra of many young people. In fact, these influences seriously shook the pillars of political power and provoked an extreme reaction from the governments of the United States and many European countries in an effort to limit them. Of course, there were numerous reasons that can be invoked to explain why these countercultural attitudes arose when they did, but it is certainly clear that they were greatly influenced by Eastern religious thinking. There is a tendency today to think about the mass cultural movements of the 1960s as "wishy-washy," narcissistic, and out of touch with political reality, the thinking on display being nothing more than the ravings of drug-addled brains—but nothing could be further from the truth. Many of these people did truly believe in peace and nonviolence, and what is more, they practiced it.

Naturally, as with every other cultural attitude, leaders were required to bring the principle of nonviolence to the public's attention in a zeitgeist-appropriate manner. Gandhi and King did this through their Hindu and Christian associations, but there were others we should note as well. Aldous Huxley, for example, was a member of a family that achieved a great deal in the nineteenth and twentieth centuries. He was the grandson of Thomas Huxley, "Darwin's bulldog"; his brother Julian Huxley was famous as a biologist and anthropologist; and his half-brother Andrew Huxley won the Nobel Prize for his work in elucidating the ionic basis of the action potential and was certainly one of the greatest physiologists in history. Like his brothers, Aldous Huxley wanted to be a scientist as a young boy, but an accident ruined his eyesight: he was practically blind and so was "Eyeless in Gaza" as well as everywhere else, reducing his effectiveness in the laboratory setting.[17] Huxley turned his vision inward and discovered a world of scientific imagination instead. He also discovered Eastern philosophy and its ancient roots—in particular, the idea of perennialism or universalism, which believes that there is a *Prisca Theologia*, a set of common human ideals, including universal love, peace, and nonviolence, that was originally transmitted to mankind in the distant past and exists as the common basis of all the world's religions. Unfortunately, according to some perennialist writers such as Rene Guenon, many of these important principles have now gone missing, particularly in

Western countries, resulting in our current lapsarian state, whereas they have been maintained in many Eastern cultures. Renaissance philosophers like Marsilio Ficino, who discovered indications of perennial thinking when translating newly discovered manuscripts in fifteenth-century Florence, attempted to fuse these ideas with Christian thinking. This clearly shows that attempts at reintroducing Eastern non-Christian thinking back into the West are nothing new. Huxley wrote his book *The Perennial Philosophy* to describe this idea, quoting widely from Eastern and Western religious texts that supported this principle. Additionally, Huxley's ideas about Eastern philosophy as well as his ideas about psychedelic drugs found fertile soil in the cultural milieu of the 1960s. Putting everything together, he showed how the use of psychedelic drugs could provide a shortcut to transcendence. *The Perennial Philosophy* and *The Doors of Perception* were the bookends of every hippy bookshelf.

It is likely that some of these ideas, particularly the concept of nonviolence toward all creatures, would resonate strongly today just as they did more than fifty years ago, and adherence to organized religions or psychedelic drug taking is not required; nonviolence can be an entirely secular principle. As we have discussed (see Chapter 7), in the United States in 2020, pets were present in around 70% of households, or eighty-five million families.[18] Pets and their care constituted an over $100 billion industry. Think about it; those are huge numbers! No politician receives support from 70% of the population, but animals receive this support. This is a source of untapped power that could be mobilized—the idea of nonviolence toward animals, a perennial idea that comes to us from deep history, would find a secure home in the hearts of many people. Nonviolence for this particular purpose has never been tapped; Gandhi and King generally had other aims in mind. Of course, there have been attempts to promote related issues such as environmentalism, something that can be viewed as a response to violence against the living planet Earth, and there is no reason these issues could not be combined politically. Environmental political groups such as the Green Party have failed to achieve large-scale popular support perhaps because the concept of the planet Earth is too distant from most people's personal experience; but their animal companions are not. Politicians need to understand that the idea of nonviolence toward all creatures could be a powerful motivator for many voters and could be mobilized to great effect: then perhaps we would see changes in the laws that protected animals in laboratories and elsewhere.

The Welfare of Animals and the Law

All of this is not to say that there are no efforts to reform animal research these days. Indeed, there are many organizations whose goal it is to reform the treatment of animals. These groups have tried to popularize different versions of the arguments discussed above, that animals should be respected as individuals, that they are the "subject of a life," that animals are entitled to "rights" under the law, and that animals do not constitute "property." Moreover, animal welfare[19] organizations are not just concerned with the use of animals as research subjects but cover a multitude of issues such as factory farming, vegetarianism/veganism, the use of animals in products like clothes, the status of zoos, the use of animals in sports, the status of animals as companions and pets, animal conservation, and so on.

These groups are privately organized and funded and are not usually endorsed by any of the major political parties in most countries. Rather, they act as pressure groups that try to persuade the major political parties to change their policies and ultimately the law. Some of these organizations have become very well known and attract a good deal of funding, celebrity support, and publicity. Most people will have heard of PETA (People for the Ethical Treatment of Animals), for example. PETA was founded in 1980 by a small group of individuals motivated by their own experiences as well as the writings of people like Singer. Initially the group was thought to be extremely radical as it would use tactics that were considered unusual for animal welfare groups at the time. In 1981, members of PETA infiltrated a laboratory performing monkey experiments at the Institute of Behavioral Research in Silver Spring, Maryland, and took pictures of the appalling treatment the animals were subjected to.[20] This led to a police raid and conviction of the lead scientist (later overturned) on animal abuse charges. The case garnered PETA a good deal of publicity and subsequent support, and it went on to become the archetype of a new kind of animal activism. In retrospect, the kinds of tactics used by PETA and other groups are not really new. They are reminiscent of the Brown Dog affair, which began by activists infiltrating a medical student class and recording what went on there as a basis for legal action (see Chapter 7). In a more contemporary context, PETA's actions have served to highlight the widespread latent support for animal rights among the public at large. Some forty years later, PETA is no longer regarded as the radical fringe of animal activism. The animal welfare movement has splintered into a wide range of different factions and approaches, some with even anarchist associations that will literally "stop at nothing" to further their beliefs and continue to use tactics such as the liberation of animals from laboratories or direct threats to the lives of researchers and their families. These things are no longer front-page news. Groups like PETA are now viewed as representing the center of the movement and, in fact, have been attacked by more radical animal welfare organizations as endorsing a step-by-step approach to animal welfare reform that simply makes the entire enterprise more palatable to middle-class sensibilities rather than really solving the problem. Other groups favor a more "abolitionist" approach, which condemns the use of sentient animals for research or any other purpose.[21] Whatever their individual aims and methods, animal welfare groups have clearly tapped into feelings that are widely held by members of the public and have put those who conduct animal experiments on notice that they can no longer expect to do whatever they want free from public scrutiny, something that has resulted in medical research becoming even less transparent as scientists try to shield themselves from what is going on. Moreover, it is also clear that the constant pressure applied by animal welfare groups has produced practical results, even though these are piecemeal, in terms of changes to the laws or rules that mandate what kinds of research can be carried out.

Unfortunately, the laws that currently cover animal welfare are clearly inadequate and, in some cases, ridiculous. Let us consider the situation in the United States, which is by far the world's largest consumer of animals for experimental research. As we have seen (see Chapter 7), the first laws that addressed the use of animals were passed in the United Kingdom in the nineteenth century and then soon afterward by other European countries, but not the United States. The United States certainly had antivivisectionist and related pressure groups that were founded in the nineteenth century, but their efforts never resulted in federal legislation that protected animals. This did not occur until passage of the Animal Welfare Act (AWA) in 1966, followed by its numerous amendments over the years.[22] Of course, the AWA is not the whole story because different states also provide a further degree of animal protection in many instances.[23] Nevertheless, because the AWA represents the federal position on animals, it sets the tone for the official policy of the United States when considering these issues and provides what is considered to be a basically acceptable standard for animal welfare. In 1970, the law was amended to specifically indicate that all warm-blooded animals should be covered, and the law was to be enforced by the US Department of Agriculture (USDA). That sounds fine. There is, however, one tiny but important caveat. According to the law, mice, rats, and birds are not defined as animals.[24] Research groups and similar stakeholders opposed the AWA, contending that rats, mice, and birds get ample protection from institutional IACUCs (Institutional Animal Care and Use Committee, see below) and don't need to be covered by the AWA. It was also claimed that enforcing the law would just be too expensive. I'm not kidding about this. One can read, for example, that, "Enacted January 23, 2002, Title X, Subtitle D of the Farm Security and Rural Investment Act, changed the definition of 'animal' in the Animal Welfare Act, specifically excluding birds, rats of the genus Rattus, and mice of the genus Mus, bred for use in research." Considering that over 99% of the approximately 111 million animals used in research in the United States every year are mice and rats,[25] this does seem to be a little bit of a problem. What, then, are mice, rats, and birds supposed to be if they are not warm-blooded animals? Plants, perhaps? As one might imagine, the fact that these creatures are not covered by the AWA does limit its effectiveness in regulating experiments that use them—and that is almost all experiments. So how did this piece of ridiculous nonsense come about? How is such a thing possible? Why is the United States one of the only Western countries in the world that does not have laws that protect rats, mice, and birds that are subjected to research and testing? Of course, as is usually the case, one needs to follow the money. Genuine scientific issues aside, one should realize that the entire research enterprise is a gigantic multibillion-dollar affair. Universities receive huge amounts of research funding for carrying out these experiments, the pharmaceutical industry uses animals for many purposes and doesn't want to completely change its well-entrenched research habits, large companies exist that specifically breed animals for research purposes, other companies produce equipment specifically for animal research applications, and on and on.

These vast commercial enterprises have spent gigantic amounts of money in supporting groups that lobby politicians to make sure that nothing interferes with their profits or cash flow potential. This has nothing to do with animal welfare. Sadly, it is just about money and is part of the alarming trend these days in which universities have just become gigantic self-serving corporations that have abandoned any role they might once have had in promoting academic excellence or ethical behavior. As a result, we are left with a piece of legislative nonsense like the AWA. Newer approaches for persuading politicians that the future lies in the support of animal welfare and not in animal abuse are sorely needed. There are certainly millions of people who would be receptive to such approaches.

There have been some attempts to do this. In 1993, congress passed the NIH Revitalization Act aimed at modernizing many of the outdated policies and regulations carried out under the sponsorship of the NIH. This law included a substantial section addressing the growing need and opportunities for replacing animals in research. The act called upon the NIH "to conduct or support research into methods of biomedical research and experimentation that do not require the use of animals" and "for training scientists in the use of such non-animal methods that have been found to be valid and reliable," as well as "encouraging the acceptance by the scientific community of such methods that have been found to be valid and reliable."

More recently two congressmen from Florida, a Democrat and a Republican, have drafted a bill aimed at updating animal welfare policies in research.[26] The bill will "establish the National Center for Alternatives to Animals in Research and Testing (Center) under the National Institutes of Health (NIH)," which "will be dedicated to increasing transparency and understanding regarding the use of animals in medical research and testing to ultimately reduce the number of animals utilized in such practices" and "allow the NIH to develop, fund, and execute a plan to record an accurate account of animals used in testing and research and to incentivize the use of non-animal methods by educating and training scientists to utilize alternative 'human relevant' methods." This sounds sensible, of course, and is in keeping with current moves to use human-based experimental paradigms, but whether this attempt at legislation will succeed or fail or be diluted by other interests remains to be seen. Nevertheless, the very fact that this kind of legislation now seems politically feasible indicates that scientists are under increasing pressure to stop animal experimentation and look for other methods.

Fortunately, in addition to federal law, animal protection legislation in the United States exists at the state level, and recently, in keeping with the public's increasing interest in animal welfare, there have been a host of new laws passed that regulate things like the testing of cosmetics, the selling of fur, and the declawing of cats. However, these laws don't generally concern the use of animals in laboratory research. Here, protection is supposed to be the purview of the institutional IACUC. This is a committee formed by members of the university or other research institution usually consisting of faculty members and veterinarians whose job is to review internal research proposals requesting to use laboratory animals in experiments. Members of the faculty who wish to carry out such experiments must submit a detailed account describing what they want to do, including the types and numbers of animals they wish to use, the precise details of the experiments they wish to carry out, the type of anesthesia involved if necessary, and other pertinent details. This is reviewed by the IACUC, which will grant permission or request revisions to be made. In extremis, they may even reject the proposed experiments. Permission is based on adhering to rules specified by a manual (*The Guide for the Care and Use of Laboratory Animals*), which is supported by the NIH, by far the largest funding body for university research in the United States. In other words, animal care is basically an "honor" system overseen by the research institutions themselves rather than being under the supervision of the USDA or other federal agency. Clearly, one might imagine that there could be a conflict of interest in having universities oversee their own animal experimentation: a Turkish proverb states that "You don't get the cat to carry home the cream." But in truth, there isn't a lot of evidence that this is the case. In my limited experience, the IACUCs at universities are run by unusually dedicated individuals who spend a great deal of time reviewing all these applications and are rigorous in their adherence to the rules provided by "the guide." Nevertheless, if animal experiments are going to take place, one would feel more confident if the government were directly involved in policing the entire system, and in other countries that is what happens. The inspectors who oversee animal experimentation are civil servants who work for the government, and the laws that govern these procedures are the laws of the country.

Indeed, many countries have much stricter laws governing animal experimentation than the United States. Some countries are particularly noteworthy, such as Switzerland.[27] Switzerland has a strong animal welfare history and some of the strictest animal welfare laws in the world, which, in recent times, have become progressively more restrictive, resulting in a substantial decline in the number of animals used in research. At this time, permission to carry out animal experimentation in Switzerland is based on a utilitarian calculus. Before conducting experiments with animals, researchers must obtain a permit from the veterinary office in their local canton, which makes decisions based on the recommendation of a regional animal experimentation commission—a group that includes representatives from both research institutions and animal protection organizations. This process requires scientists to submit detailed descriptions of their proposed protocols and to demonstrate that the societal benefits gained from their research outweigh the suffering of their subjects. However, the relative value of animal experiments in the context of animal suffering is much more weighted toward animal welfare in Switzerland than in many other countries, and scientists have to jump through a number of difficult hoops to get their work approved. The emphasis on this utilitarian analysis stems from the 1992 referendum in Switzerland in which the public voted to include the protection of animal dignity in the country's constitution and from the subsequent implementation of several changes to animal protection legislation over the last decade. It is felt that, even in the face of recent events such as the COVID-19 epidemic, the Swiss animal welfare laws are destined to become more restrictive as time goes on.

As opposed to countries like the United States, Switzerland has a much more agile system for potentially allowing popular opinion to influence its constitution. This takes place through the calling of referenda, which can be initiated by citizens who collect over 100,000 signatures supporting a proposal over a period of eighteen months. In 2018, 124,000 signatures were collected to support a proposal that animal experimentation was the equivalent of torture and was therefore a criminal offense and that suggested banning the majority of all animal research. This initiative was launched in 2017 by citizens in St. Gallen in eastern Switzerland and has the support of eighty groups and businesses as well as the leftwing Social Democratic and Green Parties. The government and parliament still need to discuss the proposal before it is put to a nationwide vote, expected in 2022. Switzerland has one of the most powerful pharmaceutical industry lobbies in the world and numerous outstanding research universities, so there has been plenty of pushback against the referendum. Even so, it has a chance of success (but see note [28]). If we presume, as discussed above, that support for animals in the United States is as widespread as it is in Switzerland, this serves to illustrate the political potential of people in the United States who care about animals even in the face of powerful lobbies.

Some countries have very little in terms of legal protection for animals. Here we should note China, in particular, where an enormous amount of cutting-edge biomedical research takes place these days. Because of the growing condemnation of animal research in Western countries,[29] we have seen a reduction in the use of animals like monkeys in recent years in Europe and the United States. Unfortunately, scientists who want to carry out research on monkeys or perform studies that would be considered unethical in the United States or Western Europe can still collaborate with their colleagues in China and so get around problems associated with performing such work in their native countries. Recently, for example, an elderly Chinese scientist set up what basically amounts to a concentration camp for monkeys in China where, he says, scientists will be able to use things like the latest genetic techniques to produce gene-edited monkeys as models for human diseases.[30] To illustrate the point, he published a paper showing how one could use these techniques to edit a gene that destroyed sleep rhythms in monkeys, presumably because he believed that disrupting sleep in humans was an important and relevant therapeutic goal.[31] As might be expected, several renegade Western scientists have jumped at the opportunity of working at this new institute. One senior German researcher who uses monkeys and whose lab was recently closed because of animal abuse problems appears to be joining up, certainly not a promising trend and something that illustrates the unfortunate consequences of setting up an institute of this type in a country where animals have little legal protection.[32] Nevertheless, it would be wrong to think that everybody in China speaks with one voice when it comes to animal experimentation. A recent report involving a particularly worrying group of experiments carried out on mice was publicly criticized by Chinese ethicists.[33] Animal welfare needs to become a worldwide issue, and countries like the United States that wield a great deal of power need to take a stand in promoting nonviolence to animals wherever it occurs.

Whatever the legal status of animal experimentation in the United States and other countries, scientists are already supposed to be obliged to stop working on animals whenever possible. Why is this? In 1959, two British investigators, W. M. S. Russell and R. L. Burch, published a book entitled *The Principles of Humane Experimental Technique*. One suggestion the authors made was the adoption of the "3Rs," three rules for enhancing the humane use of animals in science and animal welfare:[34]

Replacement: methods that avoid or replace the use of animals in research
Reduction: use of methods that enable researchers to obtain comparable levels of information from fewer animals, or to obtain more information from the same number of animals
Refinement: use of methods that alleviate or minimize potential pain, suffering, or distress and enhance animal welfare for the animals used

Suggestions such as these hoped to ensure that scientists constantly updated their attempts to reduce the use of animals and look for alternative techniques. These recommendations have become widely accepted by the public and governments throughout the world. The NIH Revitalization Act of 1993 specifically mandated the development and dissemination of the 3Rs within the scientific community. Indeed, these suggestions have become very influential in some circles as guiding principles on how we are supposed to approach animal experimentation. People who work in organizations that support reforms in animal welfare and experimentation are certainly aware of these principles. They write about them a great deal and discuss them widely. You would therefore imagine that they would be influential in the actual practice of animal-based research. However, there is one unfortunate problem with all of this—the scientists who conduct the research appear to have never heard of the 3Rs! What do I mean by this? After all, as we have seen, there are animal welfare committees who have to review all the experiments that are carried out in universities or industry, and there are laws that dictate what you can and cannot do. It is very likely that the veterinarians and faculty members who serve on these committees have heard of the 3Rs and take them seriously. However, scientists outside this select group appear to be entirely ignorant of the 3Rs. In over forty years as a scientist, I have never heard a single investigator bring them up in everyday discourse. This is hardly surprising considering that there are no efforts whatsoever to encourage researchers to discuss these things. Scientists talk about issues concerning science, but they never discuss the ethical appropriateness of animal research. Perhaps it isn't the scientists themselves who should take the blame for this. Research universities make no efforts to incorporate issues like these into the fabric of scientific life (see Chapter 9). Basic research scientists have lectures, seminars, and symposia on the latest in cancer research or other important fields of science but never about ethical considerations that cover issues like animal usage; as a matter of fact, they don't have lectures on ethical issues in science of any kind. When students enter graduate school, it is likely that they have a course in "research ethics," and it is possible that things like the 3Rs get a mention in passing. They are obliged to take such courses by funding agencies, but their hearts are not in it. There are no efforts to maintain students' interest in these matters once the administrative box has been checked, and the result of this is that students aren't interested in these topics, or at least aren't encouraged to be interested in them.

In many respects, scientists are very conservative. If you have spent many years developing a genetic system that allows you to give a mouse a horrible disease, you are not going to welcome somebody knocking on your door telling you that what you are doing is out of date and unethical and that you should be doing something else. Moreover, the scientists who determine how research funds are distributed by the government and other organizations are mostly in the same boat and are not inclined to raise these issues. But of course, things can change, and in the end, if the pressures are strong enough, they will. Ten years ago, if you were conducting basic research using mice or other animals, unless you were specifically investigating some aspect of female health and physiology, all your experiments were performed on male mice. It was thought that female mice were just "too variable" to allow rigorous studies to be easily performed—*la donna e mobile!* How could you produce beautiful data sets with straight lines and high degrees of statistical significance with all that variability? Of course, the way males and females respond to diseases and to medication in the real world differs in many important respects, so therapies that were developed that ignored matters relating to gender often didn't work as well as they might, for obvious reasons. For example, males and females experience pain very differently and have different responses to pain medications—obviously a hugely important consideration in the practice of medicine.[35] A recent publication has highlighted the numerous differences in gene expression that occur in the nerve pain pathways of men and women.[36] Eventually, mainly owing to the voices of female scientists, attempts were made to rectify this situation, and these days, if you don't explain why you aren't considering things like gender in your studies, you can basically forget it. As Nietzsche wrote in *Beyond Good and Evil,* "Supposing that truth is a woman—what then?"

One result of this is that the science that we do is better. It is more inclusive and more relevant, and sooner or later, public pressure and the realization that human-based experimental systems result in better, more meaningful data will lead to similar changes with respect to animal research. Universities will start to feel pressure where it hurts them most—in their ability to attract money. Public pressure has already led to legal changes in some countries such as Switzerland that have stringent laws restricting animal usage, things that universities in that country are now obliged to take very seriously. I think that such a scenario is what is in store for the scientific establishments in the United States once there is an appropriate political outlet for public opinion in support of animal welfare.

Scientists could be clever about all of this if they wanted to be. The scientists of the future will have to be, and they should be. What if scientists, universities, and pharmaceutical companies embraced matters concerning animal experimentation? Take Johns Hopkins University's medical school for example, which in most people's estimation is the best and most innovative in the United States. The Johns Hopkins Center for Alternatives to Animal Testing (CAAT) was founded in 1981.[37] This move was made in response to pressure from the public owing to revelations about the use of animals in the cosmetics industry. One goal for CAAT is to determine how scientists can apply the 3Rs to their work. CAAT provides a basis for the widespread discussion of these issues within Johns Hopkins University and beyond. It teaches courses and awards degrees. Surely this is what every research institution should be doing? Every major research institution should have the equivalent of CAAT where issues relating to animal research are discussed and active efforts are made to really reduce these practices. Clearly, there is now a pathway toward performing outstanding medical and biological research that no longer uses animals, and scientists need to actively engage in this or they will be left behind.

Although we may recognize the fact that the technological underpinnings of biology are moving away from animal experimentation and embracing experimental paradigms that are based on humans, this will not happen overnight. Most laboratories are still run by old scientists who are set in their ways and will not or cannot see into the future—but these people are on the way out. A new type of scientist will soon arise, someone who is able to see clearly how to direct experimental research with the aim of solving the problems that afflict humans and is able to understand the ethical and cultural consequences of what they are doing. The Enlightenment that began in the seventeenth century freed science from its metaphysical associations and provided reliable methods for the creation of new knowledge. Now, the Second Enlightenment of the twenty-first century will teach scientists what they must do to gather this new knowledge in the context of an ethical landscape that is consistent with the ideals of a world at peace with itself and all its creatures.

If the scientists of today look at themselves in the mirror of morality, all they will see is nothing. They have donned the Ring of Gyges and have secreted themselves far from public scrutiny by becoming invisible; but as we all know, rings of invisibility have a way of slipping off, leaving their wearers exposed to the stark light of public opinion from which there is no escape. In contrast, the scientists of the future will be able to look in the same mirror and recognize themselves as morally and ethically superior, engaged with the world and all its creatures, ready to carry science forward into a better future.

Notes

1. Bernard E. Rollin, "Animal Research: A Moral Science: Talking Point on the Use of Animals in Scientific Research," *EMBO Reports* 8, no. 6 (June 2007): 521–525, doi:10.1038/sj.embor.7400996.
2. I had heard this from Katie Watson, JD, a colleague of mine in the Center for Bioethics and Medical Humanities at Northwestern University.
3. Peter Singer, *Animal Liberation: The Definitive Classic of the Animal Movement*, updated ed. (New York: Ecco Book/Harper Perennial, 2009).
4. Anita Guerrini, *Experimenting with Humans and Animals: From Galen to Animal Rights*, Johns Hopkins Introductory Studies in the History of Science (Baltimore, MD: Johns Hopkins University Press, 2003), 114–136.
5. César Muñoz-Fontela et al., "Animal Models for COVID-19," *Nature* 586, no. 7830 (October 2020): 509–515, doi:10.1038/s41586-020-2787-6.
6. Richard Van Noorden, "The Science That's Never Been Cited," *Nature* 552, no. 7684 (December 2017): 162–164, doi:10.1038/d41586-017-08404-0; "News at a Glance," *Science* 370, no. 6515 (October 2020): 386–387, doi:10.1126/science.370.6515.386; Mira van der Naald et al., "Publication Rate in Preclinical Research: A Plea for Preregistration," *BMJ Open Science* 4, no. 1 (August 2020): e100051, doi:10.1136/bmjos-2019-100051.
7. "News at a Glance"; van der Naald et al., "Publication Rate in Preclinical Research."
8. Tom Regan, *The Case for Animal Rights* (Berkeley: University of California Press, 2004).
9. Lawrence Wright, "The Elephant in the Courtroom," *New Yorker*, February 2022, https://www.newyorker.com/magazine/2022/03/07/the-elephant-in-the-courtroom.
10. Natalie Evans, "Agency and Autonomy: A New Direction for Animal Ethics" (PhD thesis, UWSpace, 2014), http://hdl.handle.net/10012/8158; Natalie Thomas, *Animal Ethics and the Autonomous Animal Self* (London: Palgrave Macmillan UK, 2016), doi:10.1057/978-1-137-58685-8.
11. Chad Wellmon, *Organizing Enlightenment: Information Overload and the Invention of the Modern Research University* (Baltimore, MD: Johns Hopkins University Press, 2015).
12. Phillip A. Lecso, "To Do No Harm: A Buddhist View on Animal Use in Research," *Journal of Religion & Health* 27, no. 4 (December 1988): 307–312, doi:10.1007/BF01533198.
13. Peter Abelsen, "Schopenhauer and Buddhism," *Philosophy East and West* 43, no. 2 (April 1993): 255, doi:10.2307/1399616; Christopher Ketcham, "Schopenhauer and Buddhism: Soulless Continuity," *Journal of Animal Ethics* 8, no. 1 (April 2018): 12–25, doi:10.5406/janimalethics.8.1.0012.
14. azquotes, "Arthur Schopenhauer Quotes About Animals," n.d., accessed July 19, 2022, https://www.azquotes.com/author/13133-Arthur_Schopenhauer/tag/animal; goodreads, "The Basis of Morality Quotes," n.d., accessed July 19, 2022, https://www.goodreads.com/work/quotes/113207-the-basis-of-morality.

15. Derrick Everett, "Vegetarianism and Antivivisection," June 2022, https://www.monsalvat.no/vegetarianism.htm.

16. Ira Chernus, "The Role of Religions in Promoting Non-Violence," *Diogenes* 61, no. 3–4 (August 2014): 46–58, doi:10.1177/0392192116648595.

17. Clive James, "Aldous Huxley, Short of Sight," *New Yorker*, March 2003, https://www.newyorker.com/magazine/2003/03/17/aldous-huxley-short-of-sight.

18. "Facts + Statistics: Pet Ownership and Insurance," Insurance Information Institute, n.d., https://www.iii.org/fact-statistic/facts-statistics-pet-ownership-and-insurance.

19. Arguably animal welfare, animal rights, and animal liberation don't necessarily mean the same thing but have become somewhat conflated with each other in common discourse and so I use the term "animal welfare" to apply to all of them.

20. "Silver Spring Monkeys," *Wikipedia,* n.d., accessed July 19, 2022, https://en.wikipedia.org/wiki/Silver_Spring_monkeys#Pacheco's_description_of_the_laboratory.

21. "Animal Rights: The Abolitionist Approach," n.d., https://www.abolitionistapproach.com/about/mission-statement/.

22. "Laws That Protect Animals," Animal Legal Defense Fund, n.d., https://aldf.org/article/laws-that-protect-animals/.

23. Natasha Daly, "States across U.S. Are Taking Bold Steps toward Protecting Animals," *National Geographic*, July 2019, https://www.nationalgeographic.com/animals/article/first-time-animal-welfare-laws-in-us-states.

24. "Rats, Mice, and Birds," Animal Welfare Institute, n.d., https://awionline.org/content/rats-mice-birds.

25. Larry Carbone, "Estimating Mouse and Rat Use in American Laboratories by Extrapolation from Animal Welfare Act-Regulated Species," *Scientific Reports* 11, no. 1 (December 2021): 493, doi:10.1038/s41598-020-79961-0.

26. Aysha Akhtar and Barbara Stagno, "This New Bill Could Phase Out Animal Testing for Good," *Sentient Media*, June 2021, https://sentientmedia.org/this-new-bill-could-phase-out-animal-testing-for-good/.

27. Erin Evans, "Constitutional Inclusion of Animal Rights in Germany and Switzerland: How Did Animal Protection Become an Issue of National Importance?," *Society & Animals* 18, no. 3 (2010): 231–250, doi:10.1163/156853010X510762; Diana Kwon, "Swiss Researchers Struggle to Get Animal Experiments Approved," *The Scientist*, January 2019, https://www.the-scientist.com/news-opinion/swiss-researchers-struggle-to-get-animal-experiments-approved—65293; "Switzerland to Vote against Animal Experimentation," *Le News*, April 2019, https://lenews.ch/2019/04/16/switzerland-to-vote-against-animal-experimentation/.

28. Fabienne Crettaz von Roten and Martin W. Bauer, "Votes Show Swiss Public Still Supports Some Animal Research," *Nature* 603, no. 7902 (March 2022): 578–578, doi:10.1038/d41586-022-00785-1.

29. David Grimm, Harvard Studies on Infant Monkeys Draw Fire," *Science* 378: 341–342.

30. Dennis Normile, "China's Big Brain Project Is Finally Gathering Steam," *Science* 377: 1368–1369.

31. Peiyuan Qiu et al., "BMAL1 Knockout Macaque Monkeys Display Reduced Sleep and Psychiatric Disorders," *National Science Review* 6, no. 1 (January 2019): 87–100, doi:10.1093/nsr/nwz002.

32. Gretchen Vogel, "Monkey Facility in China Lures Neuroscientist," *Science* 367, no. 6477 (January 2020): 496–497, doi:10.1126/science.367.6477.496-b.

33. Smriti Mallapaty, "'Pregnant' Male Rat Study Kindles Bioethical Debate in China," *Nature* 595, no. 7868 (July 2021): 481, doi:10.1038/d41586-021-01885-0; Kathleen McLaughlin, China Finally Setting Guidelines for Treating Lab Animals: Scientists Hope Standards Will Benefit Both Animals and Research Efforts," March 21, 2016.

34. "The Three Rs," *Wikipedia*, July 2022, https://en.wikipedia.org/wiki/The_three_Rs; Robert C. Hubrecht and Elizabeth Carter, "The 3Rs and Humane Experimental Technique: Implementing Change," *Animals* 9, no. 10 (September 2019): 754, doi:10.3390/ani9100754.

35. Amber Dance, "Why the Sexes Don't Feel Pain the Same Way," *Nature* 567, no. 7749 (March 2019): 448–450, doi:10.1038/d41586-019-00895-3.

36. Jennifer Mecklenburg et al., "Transcriptomic Sex Differences in Sensory Neuronal Populations of Mice," *Scientific Reports* 10, no. 1 (December 2020): 15278, doi:10.1038/s41598-020-72285-z.

37. "Johns Hopkins University Center for Alternatives to Animal Testing," n.d., accessed July 19, 2022, https://caat.jhsph.edu/.

9

The Cloud Cap'd Towers

In the whole human race / there is no greater criminal / than a man without money.

— *The Rise and Fall of the City of Mahagonny*, Kurt Weil/Berthold Brecht, 1930

The dismal specter of cruelty to animals carried out in the name of science is rapidly becoming a thing of the past. We won't miss it. As we have seen, there are numerous reasons we can do without animal-based research nowadays. The argument that it is essential is rarely true anymore. This conclusion is based on purely scientific reasoning—the growing realization that animal experimentation has been superseded by human-based technologies that are far better placed to solve human-related problems. As we move forward and embrace a more effective, kinder, and more ethical type of research, we need to reimagine the scientists of the future who will lead the transition and run the laboratories where research will be planned and carried out. In the future, if scientists must consider killing animals when performing experiments, it should be inevitable that they contemplate their actions very carefully. After all, they are depriving creatures of their lives. But scientists aren't trained to do that these days. There are few serious attempts to get them to think ethically or broadly about what they are doing. Now, as the character of biomedical research is changing, we have an opportunity to educate the scientists of the future so that they think about their work not only as science but also in a broader context that includes the world of influences outside the hermetically sealed bubble where scientists currently live and work. The bubble needs to burst. We need a revolution in the training of the scientists of the future. The revolution will have to take place within a conservative and sometimes even hostile environment—the modern research university.

The Rise and Fall of Animal Experimentation. Richard J. Miller, Oxford University Press. © Richard J. Miller 2023.
DOI: 10.1093/oso/9780197665756.003.0009

The University Tradition

Modern universities certainly look very nice—more and more beautiful buildings, designed, of course, by world-renowned architects. But are the people who work in these buildings productive in the best sense of the word, their laboratories pouring out reams of useful data? Unfortunately, as we have discussed, it's frequently just the opposite, the piles of information being produced having very little real value beyond supporting the careers of scientists. The current situation where billions of animals are routinely put to death for no real purpose is just one component of a crisis affecting the entire structure and goals of research universities. Universities have drifted far away from any origins they may once have had as pure bastions of knowledge. Rather, they have become giant corporations whose major goal is the accumulation of cash. Naturally, scientists have been caught up in all of this and have adapted their modus operandi accordingly, mostly to their own detriment. Scientists are now caught in a vice, squeezed between the financial demands of their universities and the expectations of a scientific culture that has become increasingly dissociated from the real world and whose view of success often has little to do with scientific truth. The universities and the culture of science now work together to such malevolent effect that scientists have less and less time to devote to their genuine scientific interests and critical evaluation of their work as they try to keep their heads above water and prevent themselves from drowning in a sea of red ink. According to one review:[1]

> Over the last 50 years, we argue that incentives for academic scientists have become increasingly perverse in terms of competition for research funding, development of quantitative metrics to measure performance, and a changing business model for higher education itself. Furthermore, decreased discretionary funding at the federal and state level is creating a hypercompetitive environment between government agencies, for scientists in these agencies, and for academics seeking funding from all sources—the combination of perverse incentives and decreased funding increases pressures that can lead to unethical behavior. If a critical mass of scientists become untrustworthy, a tipping point is possible in which the scientific enterprise itself becomes inherently corrupt and public trust is lost, risking a new dark age with devastating consequences to humanity. Academia and federal agencies should better support science as

a public good, and incentivize altruistic and ethical outcomes, while de-emphasizing output.

This has to change, and universities need to embrace an enlightened approach to education that is commensurate with a new type of wide-ranging ethical science.

Universities can certainly change. Indeed, the university is a particularly protean institution that, throughout history, has adapted itself to the world around it by taking many forms. It's an interesting story and it is worth spending a moment to consider how universities developed their contemporary characteristics.[2] Perhaps the earliest university, at least as far as medicine was concerned, was Salerno, whose curriculum was established under the influence of Constantine the African who brought translations of the classical works of Hippocrates, Aristotle, and Galen with him from Arab North Africa across the Mediterranean to southern Italy in the eleventh century. However, the first institutions with comprehensive curricula included Bologna, Oxford, Cambridge, and Paris, all founded in the eleventh and twelfth centuries. In Paris, in particular, charismatic teachers like Peter Abelard, Albertus Magnus, and his pupil Thomas Aquinas set out to reconcile Christian teaching with the recently rediscovered writings of Aristotle, the basic tenets of scholasticism. These were exciting new ideas that attracted willing students to these institutions from throughout Europe. Universities eventually carved out a niche for themselves as a "third estate," an independent polity that at various times could ally itself with either the church or state depending on the situation.

The modern "research university," which included professional scientists among its faculty,[3] developed primarily in Germany in the nineteenth century. Its purpose was a combination of teaching and research rather than merely the conservation and dissemination of traditional knowledge,[4] which had been the major goal of universities for centuries. In Germany, a program to reform the university system was made the responsibility of Wilhelm von Humboldt, brother of the famous scientist and explorer Alexander von Humboldt, a project that was influenced by the writings of Romantic philosophers such as Schelling and Fichte. Von Humboldt brought an entirely new perspective to the aims of the university. In prior times, the ideal was that every individual university faculty member should strive to be an "encyclopedia," achieving complete erudition. The world consisted of various pieces of knowledge that could be collected, categorized, and taught from one generation to the next, usually by single professors teaching all subjects. This was the traditional role of the university: to know what was true and to consolidate and transmit this knowledge. But what von Humboldt suggested was that a university should also be involved in research to uncover new facts so that the totality of knowledge was always evolving and could therefore never be completely known. The university professor should be not only somebody who collected, collated, and transmitted all available knowledge but also somebody who contributed to the historical progression of knowledge through time. Universities would constitute the seat of epistemic authority, which seemed under threat as scientists attempted to cope with the unprecedented growth of new data that was being produced by the burgeoning scientific revolution of the industrial era. The new kind of professor would not be expected to teach everything but would be a specialist, characterized by the pursuit of knowledge through a focus on the particular. Furthermore, in von Humboldt's view, professors should also imbibe and transmit to their students the spirit of *Bildung,* or "self-formation," a state of moral and spiritual depth.[5] Hence, it was always the case, from its very inception, that the faculty at a research university should be involved in the moral and ethical aspects of training students in addition to pure research. Research universities don't do this nowadays; ethical training of scientists certainly wouldn't be on the curriculum. However, we should certainly ask ourselves whether in today's world such training is still desirable and whether or not something important has been lost through its omission. Von Humboldt's ideas led to the founding of his flagship institution, the University of Berlin (1810), which was envisioned as being a university that would reflect all the aims of the Enlightenment and would become a center not only for nurturing the development science and technology but also for discussing the place of science in culture in general.

The Endless Frontier

It was the period after World War II that saw the greatest expansion in higher education since the twelfth century and set the scene for the type of university training we are familiar with today. Three-quarters of all universities, even in Europe, were founded in the twentieth century, particularly since 1945, to provide highly trained professional experts. These universities embraced the concept of higher education for all rather than education just for the ruling elite, which accompanied the transition from an aristocratic to a meritocratic society. This, of course, has been a welcome development. Higher education should not discriminate between genders or races; inclusiveness should be one of its most important aims.

As universities have grown, so has the size and range of the research enterprises they foster. Biomedical research in universities these days is a gigantic multibillion-dollar affair enabled by the enormous amounts of funding these institutions receive, primarily from the federal government. Things weren't always this way. Prior to the Second World War, university research was funded by the universities themselves or perhaps through contracts with some industries for highly goal-directed projects whose results could be realistically assessed according to their practical aims. Then, during the war, the Allied and Axis powers poured money into research, and the results were impressive. From radar to antibiotics to the atomic bomb, scientists made numerous discoveries that were essential to the war effort, and many of these things have subsequently become part of modern life. In Germany the Interessengemeinschaft Farbenindustrie AG (Combined Interests of the Dye Making Industry), known as IG Farben, made many important discoveries, including antimalarial drugs like chloroquine and opiates like meperidine (Demerol), as well as synthetic rubber forged by the slave labor of prisoners at Auschwitz.

After the war, the idea that government investment in basic research could yield important results that were helpful to society in general was taken up in the United States by Vannevar Bush, who had headed the Office of Research and Development during the war and had become the scientific advisor to President Roosevelt. Bush's support for science was encapsulated in a July 1945 report entitled "Science, the Endless Frontier." Bush had a romantic vision: in addition to tying science to specific industrial contracts with clear-cut practical aims, scientists should also be supported so that they could follow their basic intuitions about Nature in general, something that would help to lay the foundations for our general understanding of the world and how it worked. This would positively influence our ability to make practical and useful discoveries in the future. Bush wrote, "Scientific progress on a broad front results from the free play of free intellects, working on subjects of their own choice, in the manner dictated by their curiosity for exploration of the unknown."[6] And as suggested by the American philosopher Simon Newcomb, "Original research in this sense was not the efforts to which we are impelled by our daily physical wants but those to which we are impelled by the purely intellectual wants of our nature."[7] Bush proved to be persuasive, the government bought into his ideas, relaxed its purse strings, and funding for basic research began to flow into the universities; and once it began it simply didn't stop. US federal funding for basic research rose from $265 million in 1953 to over $40 billion in the twenty-first century, a twentyfold increase when adjusted for inflation.[8] The lion's share of this was a huge increase for basic research at universities and colleges, which rose from $82 million to $24 billion, a more than fortyfold increase when adjusted for inflation over the same time period.[9] The flame had been ignited, and the engine of university research began to churn, resulting in the recruitment of large numbers of scientists to carry out the new research programs. Science was no longer merely a hobby for aristocrats like Robert Boyle but a profession that involved millions of people throughout the world.

After the Second World War, once the money began to arrive from the government, universities needed to look after it properly, but they were ill prepared to do this. Traditionally, faculty members had been in charge of taking care of this kind of thing, but now the flow of funds rapidly became so large that administrators needed to be hired for this specific purpose. And of course, the administrators did what they were supposed to do, and that was make efficient use of the money that they were charged to administer. These individuals knew nothing about science; their entire raison d'être was the money itself and how to make it grow. They noted, for example, that when scientists received research grants from the government, these came with some additional funds to cover "indirect costs," things that supported the infrastructure on which the research enterprise rested, things like paying the salaries of administrators. Because of this, the number of administrators grew, and their handling of research funds became more and more involved, sophisticated, and creative. For example, rather than universities paying the salaries of the faculty, why not have scientists apply for government funding to support their own salaries as well as their research? This idea proved to be a financial windfall. At some point, rather than being a nice addition to a research grant, using a grant for salary support was no longer optional, and researchers really began to feel the heat if they didn't consistently provide these funds. Soon, whatever their original intentions, faculty members found themselves strapped to the wheel of Ixion in the service of the almighty dollar, and the scientific enterprise continued to grow. Universities built new research buildings, which they then had to populate with even more scientists, who had to bring in even more research grants in order to keep pace. Universities became very competitive with one another. Magazines began to publish lists ranking universities from the best to the worst, judging them according to how much research money they were accumulating, in addition to, or perhaps rather than, any actual scientific results they produced.[10] During all of this the power of the administrators was growing, and at some point, the tail started wagging the dog. Rather than being visionaries creatively planning future research goals, the chairmen of basic science departments, stripped of any real influence, became powerless in the face of pressures emanating from the central university administration, a Kafkaesque body whose tentacles had now metastasized into controlling every conceivable aspect of university life. One thing is for sure. Nowadays science at universities isn't just research; it's business—big business.

The Modern Scientist

To increase the flow of scientists to populate all the new buildings that have resulted from the surge in government largesse since the Second World War, universities have also increased their intake of graduate and postdoctoral students. These are the people who are called upon to conduct the animal experimentation that laboratories carry out. They are also the people who will become the scientists of the future. This being the case, it is important that they are trained so that they can critically evaluate what they are doing from both a scientific and an ethical perspective.

Unfortunately, these days, training a graduate or postdoctoral student is no longer the exercise it should be. First of all, and of great importance, students frequently get poor advice about science in general; this, of course, is not their fault but the fault of their mentors. It's hardly surprising. New graduate students are assigned to faculty mentors not because they are good at mentoring but simply because they have the funds to pay the student's salary. Many of these people have little interest in training students. The result is that students are frequently assigned to laboratories that look upon them as a source of labor whose role it is to help enhance the reputation of the head of the laboratory, rather than as individuals who need careful nurturing. This results in students who have received no real advice as to how to navigate the world of science and how to contribute to it both within or outside the ivory tower of academia. These students will certainly not have received any encouragement to develop the kind of ethical compass that would include thinking about the care of animals or, indeed, other important issues relating to how science manifests itself in the world in general. Even though many of them will have begun their careers with a burning egalitarian desire to use science for the benefit of humanity, the current university atmosphere will soon disabuse them of such notions as they learn from their mentors that they need to spend the majority of their time obtaining grant funding to pay for their salaries. Sadly, this is the modern university spirit that students will inevitably imbibe.

The Crisis of Reproducibility

All of this leads to other problems as well.[11] As an illustration, let's examine one example. To be competitive for research grants, scientists need to publish their work, something that is all too easy nowadays given the huge number of scientific journals available for this purpose. This has given rise to a serious problem: the growing feeling that most of what one reads in these journals simply isn't true. By this, I don't mean that scientists simply make up their data; that is very rare indeed. Rather, there are much more systemic problems that plague the profession.

For example, what do we mean by "true," or at least what do we mean by it in a day-to-day biomedical research science sense? The validity of a scientific result is usually assessed by applying one of many statistical tests to the data that result from a series of experiments. The results of such tests provide you with a degree of confidence that the experimental results are valid and can be taken to support or dismiss a hypothesis rather than being a purely random occurrence. This is usually expressed as a "p value." If the p value < 0.05, then by convention, the data is supposed to be convincing enough to support the hypothesis in hand; if $p < 0.01$, then it is even more supportive; if $p < 0.001$, it is even more supportive; and so on. However, if the data does not reach significance of $p < 0.05$, then it is not supportive. In reality, these statistical values have assumed a much greater "significance" in normal scientific parlance such that if $p < 0.05$, the hypothesis is "true," and if $p > 0.05$, it is "not true." Science also suffers from what is called "positive bias," meaning, among other things, that journals are only interested in accepting your paper if you are going to report something that is positively significant—that is $p < 0.05$. Negative results are considered to be unimportant or at least uninteresting. But what does this mean in practice? If you have been performing experiments on a group of animals, and one animal gives you a result that is different from the rest, it is quite possible that your data won't reach statistical significance. Disaster! Not significant means not true. Not true means that you can't publish your paper. Not publishing your paper means you won't get your grant. And so on. But never fear! There are different ways in which you are allowed to curate your data, and it is possible that there will be a good "reason" you won't have to include the results from the errant animal—the animal had a bump on its head, or perhaps there is just a statistical reason. If the data is just so far from the norm, then you are "allowed" to leave it out. If you are a scientist or a graduate student whose life depends on publishing a paper, then finding a reason that will improve the validity of your data presents an almost irresistible possibility. The result of this and other factors is what has become known as a "crisis of reproducibility" in science.[12] It turns out that the results reported in the vast majority of scientific papers simply cannot be repeated by scientists in other laboratories. For example, a recent major report in the prestigious journal *E-Life* described attempts to reproduce the results of fifty-three papers in the field of cancer research over a period of several years and found that they were only successful in six cases.[13] This is because many published results aren't true in the first place, and it is easy to see why. Indeed, scientists themselves realize what is going on. A recently published survey[14] that anonymously queried 6,800 working scientists of all ranks reported extremely high numbers of scientists and students who had engaged in "dubious" research behaviors.

So, what we are faced with when presented with the scientific literature these days is a vast ocean of poorly conceived research, carried out by individuals with inadequate training, which is almost never read by anybody except the authors themselves (see Chapter 8) and, even if it is read, is unlikely to be true. And frankly, this will come as no surprise to most scientists, if they are honest about it.

Why, then, is this situation allowed to persist? The reason is that it is the result of the huge pressures that scientists find themselves under, and also because there is no real accountability within the scientific community. The reviewers of papers and grants are all in the same position as the scientists who are submitting them, and they are subject to exactly the same pressures, creating a situation where there is potentially an enormous conflict of interest and no real way of ensuring adequate standards or accountability. Scientists insist, of course, that only they have the requisite knowledge to be able to adequately critique the complex technical nature of scientific literature, and although that may be true to some extent, it is certainly also true that there are others who could assess the practical consequences of a scientific investigation. Yes, this means including educated intelligent members of the public on grant review panels whose opinions could offer a refreshing perspective beyond any discussion of scientific minutiae. As the editor-in-chief of the *Lancet*, one of the world's major medical journals, has written, "The case against science is straightforward: much of the scientific literature, perhaps half, may simply be untrue. Afflicted by studies with small sample sizes, tiny effects, invalid exploratory analyses, and flagrant conflicts of interest, together with an obsession for pursuing fashionable trends of dubious importance, science has taken a turn towards darkness."[15] One can only agree! Unfortunately, the vast mountain of scientific literature that is generated comes at a price, and the price is very high. It has been estimated that over $30 billion per year is wasted on research that is unreproducible.[16]

In spite of the fact that most scientific papers are uninteresting, unimaginative, and probably not even true, they are all accompanied by an increasing appeal to media hype. Universities are constantly at pains to describe all of the efforts of their scientists as "unprecedented" when they are addressing the media, something that might perhaps be appropriate for the Pythia at the oracle at Delphi but doesn't really describe the results in question. And, as time goes on and the competition mounts, universities and scientists must increase the hype applied to their work accordingly. Now the results are even more unprecedented or brilliant or spectacular or constitute breakthroughs of truly gargantuan proportions—the adjectival arms race is increasing at an alarming rate.

The Two Cultures Reunited

One can therefore see that there are enormous pressures on scientists that result from things that don't really have anything to do with good research and that it is hardly surprising that there is simply no time or inclination to discuss things like the ethical treatment of animals. There's just "no percentage" in it.

We should try to imagine a different type of university environment in which scientists are encouraged to think seriously about all the important aspects of their work and where the scientists of the future would receive the type of training that would help them burst out of the current scientific bubble. Such a goal may be difficult to achieve as things stand for a variety of reasons. Often, biomedical research is performed in laboratories that are next to a hospital, frequently on a completely different campus from the rest of the university. This is done so that biomedical research can understandably be directly of service to the hospital and participate in the clinical enterprise, but it also has the effect of reducing the influence of anything else such as the humanities. This is unfortunate because it is an influence that would bring something to science that it has lost—that is, "humanity" in the best sense of the word, the idea of caring and compassion about the world and all its creatures.

The humanities are concerned with matters such as ethics, philosophy, and history of science. The latter is a discipline that brings an important perspective to what scientists are doing and why they do it. Where did all this science come from? Isn't it important for scientists to understand that today's science was not just born *ex nihilo* and to be exposed to the messages of the past? Understanding the challenges that faced scientists over the centuries, the paths and blind alleys they took, how their successes and failures came about—these are inspiring stories that all working scientists need to know and understand so that they see themselves as part of the long history of scientific intellectual development. They are the inheritors of a great tradition. There are things that they need to live up to. Exposure to the humanities would make working scientists better, more thoughtful, and more compassionate. For science to engage with the humanities would be time well spent. We would do well to reconsider C. P. Snow's famous 1959 lecture bemoaning the increasing distance and lack of mutual influence between the two cultures—science and the humanities. He was undoubtably correct and it's time to reunite them.

The influence of the humanities would help scientists to take their ethical responsibilities, including their responsibilities toward animals, seriously. The ultimate challenge is to persuade the scientists of the future to behave ethically because they really want to, not because they have to—because they genuinely believe in such behavior and think that it is the correct thing to do to promote better science and a better world.

The Scientists of the Future

What is it that students learn in graduate school? At the present time, all their training is focused on scientific methodology and experimentation, and indeed, such things are certainly of great importance. But that isn't the only thing they should be learning. The choosing of a mentor for each student is a key point in the development of graduate education, and as we have seen, the selection of these mentors is based purely on financial considerations. That's it! No other qualifications are necessary. Such a situation cannot continue. Mentors for graduate students need to be carefully vetted in terms of their commitment to graduate education. We need to have assurances that they will set an appropriate standard, teaching not only how one performs experiments but also how one should think about the place of science in the world in general. Mentors need to lead by example. For a good mentor, a graduate student is not just another pair of hands to help in the laboratory but an opportunity to help in the formation of a future colleague. It is obviously important for the mentor to interact with the student to assess their scientific progress. However, in addition, the student needs to be engaged with respect to other issues that concern the practice of science in a moral dimension, including things like the appropriate use of animals. But most importantly, students learn from their mentors by watching what they do. How do they conduct research? How do they live the life of a researcher and navigate all the different issues that are important to becoming a successful scientist? What is their wisdom? There is a lot to live up to, and currently, many graduate student mentors are not living up to anything; they don't even have any real idea of what it is that they are supposed to be living up to, and so graduate students only receive an extremely impoverished education.

To this end, important changes need to be made in terms of how research universities operate so that they can enthusiastically engage in the debate about the ethics of science. One way of introducing these issues to students and faculty alike would be to organize seminars on relevant topics. Of course, research universities present lectures on scientific topics all the time, lots of them. But they need to have lectures on other issues as well, things that explore the ethical aspects of their work, science and the law, science and politics, and many other issues. Plenty of scientists today have links with the biotech and pharmaceutical industries. Many scientists think this is just another way of scoring some cash; but it's much more interesting than that, and scientists need to be aware of what they are doing. What are the benefits and problems that result from these collaborations? Biomedical researchers need to really understand that these are matters of great significance and present chances for scientists to influence events on a larger canvas.

Students love to attend lectures by "superstar" scientists. There is usually a lot of fanfare when a Nobel laureate comes to town, and students are inspired by them. But there are superstars in the humanities as well who have thought and written about scientific issues, and they need to be heard from and feted in the same way. Listening to a Pulitzer Prize winner who has written a great book about science can be an unforgettable experience. Many medical schools do have lectures, specifically for medical students, on medical ethics as applied to the relationship between doctors and patients—certainly an important topic, but that isn't what I mean. Rather, seminars should address ethical issues that apply to the practice of basic scientific research, something that is rarely discussed in research universities beyond the few rudimentary lectures which entering graduate students receive for bookkeeping purposes and which they are never encouraged to follow up. These topics should continue to be part of the curriculum though. This would also be a good place to begin a dialogue between science and the humanities departments of the university.

In addition, it would be an interesting idea if each term, a different ethical problem was selected for discussion by the entire faculty. This topic could be related to subjects about science that are in the news. Beyond this, there could be suggested readings from literature and the media that might be discussed in association with the lectures. Let us say, for example, the subjects to be discussed were the various controversies surrounding the concept of human cloning. There could be a showing of the film *Never Let Me Go* based on the novel by Kazuo Ishiguro. Perhaps in association with the humanities, there could be a performance of Caryl Churchill's play *A Number*. Alternatively, people could read the texts of the book and the play. In addition to any general discussion that might ensue, it would then be the responsibility of the head of each laboratory to meet with their lab members and discuss any interesting issues that arise and how they relate to the work that is going on in the laboratory. The entire ethics and science enterprise on each campus should be organized by a department or committee that is devoted to these topics in particular. In this way, discussions about ethical issues in science would become part and parcel of the everyday life of a working scientist, not something that is considered a complete waste of time. All these topics are interesting, and their serious consideration would improve the way we do science and the quality of the science we produce. Scientists are intelligent people, and they would certainly engage with these things if they were encouraged to do so, rewarded for doing so, and given the appropriate time.

How, then, do all these considerations ultimately apply to the use of animals? As we have seen, generally speaking, scientists don't think about this topic and just perform animal experiments because that is what they have always done, and that is what everybody else does. It's just intellectual laziness! But consider what would happen if a continuous and informed discussion about the role of animals in research was part of normal scientific training. There would be many things for scientists to think about. Most scientists don't understand how the practice of animal experimentation arose in the first place and why other cultures don't do similar things. Exposure to the history of science would explain this, and reflecting on this tradition would bring some clarity to their current practices. Yes, this is how these things arose, but is this the way they need to continue? Scientists assume that the use of animal models in biomedical research is generally the best and most effective way to proceed, but there is very little real discussion of these matters. There is very little discussion, for example, about exactly how often such results really translate to humans. Scientists will tell you that of course they do, of this they are certain, but most scientists would be surprised to know how ineffective animal studies really are in this regard. Also, as we have seen, there is the increasingly held view that the use of animals as experimental models is unnecessary because of current developments in fields such as human stem cell biology, organoid biology, in vivo imaging and transcriptomics. These are the types of issues that scientists ignore at their peril, because if they do, their work will become increasingly irrelevant as time goes on. A wider discussion of these matters would help to keep scientists current and would encourage them to start changing the way they go about their business. Perhaps Galen or William Harvey might have argued that they had no alternatives to the types of experiments they performed, but that was a long time ago. There can be no excuse for doing such things these days.

The world is changing. Some of the suggestions made here are already starting to take effect and be taken seriously by the scientific community. There is a clear movement toward the use of stem cells and organoids rather than animals, a practice that is gaining pace so quickly that it is hard to keep up with it. It's a revolution—that's for sure. The first green shoots of the Second Enlightenment are beginning to appear, and we need to water them carefully.

The use of animals in biomedical research is cruel and unethical. The use of animals in biomedical research is ineffective. The use of animals in biomedical research is unnecessary. Can we envisage a better, kinder science? Of course we can!

Vivisection in my opinion is the blackest of all the blackest crimes that man is at present committing against God and His fair creation.

—Mahatma Gandhi

Notes

1. Marc A. Edwards and Siddhartha Roy, "Academic Research in the 21st Century: Maintaining Scientific Integrity in a Climate of Perverse Incentives and Hyper-competition," *Environmental Engineering Science* 34, no. 1 (January 2017): 51–61, doi:10.1089/ees.2016.0223.

2. Harold Perkin, "History of Universities," in *International Handbook of Higher Education*, ed. James J. F. Forest and Philip G. Altbach, vol. 18, Springer International Handbooks of Education (Dordrecht: Springer Netherlands, 2006), 159–205, doi:10.1007/978-1-4020-4012-2_10; James J. F. Forest and Philip G. Altbach, eds., *International Handbook of Higher Education*, Springer International Handbooks of Education 18 (Dordrecht: Springer, 2011).

3. Sydney Ross, "*Scientist: The Story of a Word*," *Annals of Science* 18, no. 2 (June 1962): 65–85, doi:10.1080/00033796200202722.

4. Henry M. Cowles, *The Scientific Method: An Evolution of Thinking from Darwin to Dewey* (Cambridge, MA: Harvard University Press, 2020).

5. Chad Wellmon, *Organizing Enlightenment: Information Overload and the Invention of the Modern Research University* (Baltimore, MD: Johns Hopkins University Press, 2015).

6. Daniel Sarewitz, "Saving Science," *New Atlantis*, Spring/Summer 2016, https://www.thenewatlantis.com/publications/saving-science.

7. Simon Newcomb, *Exact Science in America*, vol. 119 (North American Review, 1874), 286–308.

8. Sarewitz, "Saving Science."

9. Sarewitz, "Saving Science."

10. "U.S. News Best Colleges," n.d., accessed July 18, 2022, https://www.usnews.com/best-colleges.

11. Stuart Ritchie, *Science Fictions: How Fraud, Bias, Negligence, and Hype Undermine the Search for Truth* (New York: Metropolitan Books; Henry Holt and Company, 2020).

12. Ritchie, *Science Fictions*.

13. Asher Mullard, "Half of Top Cancer Studies Fail High-Profile Reproducibility Effort," *Nature* 600, no. 7889 (December 2021): 368–369, doi:10.1038/d41586-021-03691-0.

14. Jop de Vrieze, "Large Survey Finds Questionable Research Practices Are Common," *Science* 373, no. 6552 (July 2021): 265–265, doi:10.1126/science.373.6552.265.

15. Richard Horton, "Offline: What Is Medicine's 5 Sigma?," *The Lancet* 385, no. 9976 (April 2015): 1380, doi:10.1016/S0140-6736(15)60696-1.

16. Leonard P. Freedman, Iain M. Cockburn, and Timothy S. Simcoe, "The Economics of Reproducibility in Preclinical Research," *PLOS Biology* 13, no. 6 (June 2015): e1002165, doi:10.1371/journal.pbio.1002165.

Bibliography

Adamson, Peter, and G. Fay Edwards. *Animals: A History*. Oxford University Press, 2018.

Akhtar, Aysha. *Our Symphony with Animals*. Pegasus Books, 2019.

Baldwin, A., and S. Hutton. *Platonism and the English Imagination*. Cambridge University Press, 1994.

Barnes, Jonathan. *Aristotle*. Oxford University Press, 2000.

Birke, Lynda, Arnold Arluke, and Mike Michael. *The Sacrifice: How Scientific Experiments Transform Animals and People*. Purdue University Press, 2007.

Blackmore, Susan. *Consciousness*. Oxford University Press, 2005.

Boas, Marie. *The Scientific Renaissance 1450–1630*. Fontana Books, 1962.

Bouras-Vallianatos, Petros, and Barbara Zipser. *Brill's Companion to the Reception of Galen*. Brill, 2019.

Browne, Sir Thomas. *The Major Works*. Penguin Books, 1977.

Calarco, Matthew. *Thinking through Animals*. Stanford University Press, 2015.

Calder, Louise. *Cruelty and Sentimentality: Greek Attitudes to Animals 600–300BC*. BAR International Series, 2011.

Carbone, Larry. *What Animals Want: Expertise and Advocacy in Laboratory Welfare Policy*. Oxford University Press, 2004.

Carlisle, Clare. *The Restless Life of Soren Kierkegaard*. Farrar, Straus and Giroud, 2019.

Catlos, Brian A. *Kingdoms of Faith*. Basic Books, 2018.

Clagget, M. *Greek Science in Antiquity*. Abelard-Schuman, 1955.

Clarke, Edwin, and L. S. Jacyna. *Nineteenth Century Origins of Neuroscientific Concepts*. University of California Press, 1987.

Cobb, Matthew. *The Idea of the Brain: The Past and Future of Neuroscience*. Basic Books, 2020.

Conn, Michael. *Animal Models for the Study of Human Disease*. 2nd ed. Elsevier/Academic Press, 2017.

Cooper, David E. *Animals and Misanthropy*. Routledge, 2018.

Cowles, Henry M. *The Scientific Method: An Evolution of Thinking from Darwin to Dewey*. Harvard University Press, 2020.

Crane, Susan. *Animal Encounters. Contacts and Concepts in Medieval Britain*. University of Pennsylvania Press, 2013.

de Montaigne, Michel. *The Complete Essays*. New Hall Press, 2020.

De Waal, Franz. *Mama's Last Hug: Animal Emotions and What They Tell Us about Ourselves*. WW Norton and Co., 2019.

DeGrazia, David. *Animal Rights*. Oxford University Press, 2002.

Dehaene, Stanislas. *Consciousness and the Brain: Deciphering How the Brain Codes Our Thoughts*. Viking, 2014.

Eisenman, Stephen F. *The Cry of Nature: Art and the Making of Animal Rights*. Reaktion Books, 2013.

Erdman, David V. *The Complete Poetry and Prose of William Blake*. University of California Press, 1965.

Francione, Gary L., and Anna Charlton. *Animal Rights: The Abolitionist Approach*. Exempla Press, 2015.

Fuchs, Thomas. *Mechanization of the Heart: Harvey and Descartes*. University of Rochester Press, 2001.

Garrett, Jeremy R. *The Ethics of Animal Research: Exploring the Controversy*. MIT Press, 2012.

Gluck, John P. *Voracious Science and Vulnerable Animals: A Primate Scientist's Ethical Journey*. University of Chicago Press, 2016.

Gottlieb, Anthony. *The Dream of Reason: A History of Western Philosophy from the Greeks to the Renaissance*. WW Norton and Co., 2000.

Gottlieb, Anthony. *The Dream of Enlightenment: The Rise of Modern Philosophy*. WW Norton and Co., 2016.

Gould, Stephen J. *Rocks of Ages: Science and Religion in the Fullness of Life*. Ballantine, 1999.

Greek, C. Ray, and Jean Swingle Greek. *Sacred Cows and Golden Geese: The Human Cost of Experiments on Animals*. Continuum, 2000.

Gruen, Lori. *Ethics and Animals: An Introduction*. Cambridge University Press, 2011.

Guerrini, Anita. *Experimenting with Humans and Animals: From Galen to Animal Rights*. John's Hopkins University Press, 2003.

Hannam, James. *The Genesis of Science: How the Christian Middle Ages Launched the Scientific Revolution*. Regnery Publishing, 2011.

Harris, C. R. S. *The Heart and Vascular System in Ancient Greek Medicine: From Alcmaeon to Galen*. Oxford University Press, 1973.

Harris, W. V. *Popular Medicine in Graeco-Roman Antiquity: Explorations*. Brill, 2016.

Heiser, James D. *Prisci Theologi and the Hermetic Reformation in the Fifteenth Century*. Malone, TX: Repristination Press, 2011.

Henry, John. *The Scientific Revolution and the Origins of Modern Science*. Palgrave Macmillan, 2008.

Holmes, F. L. *Claude Bernard and Animal Chemistry: The Emergence of a Scientist*. Harvard University Press, 1974.

Holmes, Richard. *The Age of Wonders: How the Romantic Generation Discovered the Beauty and Terror of Science*. Pantheon, 2008.

Hunter, Michael. *Boyle: Between God and Science*. Yale University Press, 2009.

Huxley, Robert. *The Great Naturalists*. Thames and Hudson, 2019.

Jardine, Lisa. *The Curious Life of Robert Hooke: The Man Who Measured London*. Harper Perennial, 2003.

Kinzer, Stephen. *Poisoner in Chief: Sidney Gottlieb and the CIA Search for Mind Control*. Henry Holt, 2019.

Korsgaard, Christine M. *Fellow Creatures: Our Obligations to Other Animals*. Oxford University Press, 2018.

Kramnick, Jonathan. *Paper Minds: Literature and the Ecology of Consciousness*. University of Chicago Press, 2018.

Kreiser, B. R. *Miracles, Convulsions and Ecclesiastical Politics in Early Eighteenth Century Paris*. Princeton University Press, 1978.

LeDoux, Joseph. *The Deep History of Ourselves: The Four Billion Year Story of How We Got Conscious Brains*. Viking, 2019.

Leiss, William. *The Domination of Nature*. George Braziller, 1972.

Leroi, Armand Marie. *The Lagoon: How Aristotle Invented Science*. Viking, 2014.

Lloyd, G. E. R. *Early Greek Science: Thales to Aristotle*. WW Norton and Co., 1970.

Lofting, Hugo. *The Story of Doctor Dolittle*. Yearling Publishers, [1920] 1988.

Mattern, Susan. *The Prince of Medicine: Galen in the Roman Empire*. Oxford University Press, 2013.

Maxwell-Stuart, P. G. *The Chemical Choir: A History of Alchemy*. Continuum Books, 2008.

Mayor, Adrienne. *The Poison King: The Life and Legend of Mithradates, Rome's Deadliest Enemy*. Princeton University Press, 2010.

Mitchell, Peter. *The Purple Island and Anatomy in Early Seventeenth Century Literature, Philosophy and Theology*. Fairleigh Dickinson University Press, 2007.

National Research Council. *Guide for the Care and Use of Laboratory Animals*. National Academies Press, 2011.

Newmyer, Stephen T. *Animals in Greek and Roman Thought: A Sourcebook*. Routledge, 2011.

Nietzsche, Friedrich. *Beyond Good and Evil* (trans. Helen Zimmern). SDE Classics, 2019.

Nietzsche, Friedrich. *Thus Spoke Zarathustra*. East India Publishing Co., 2021.

Olmsted, J. M. D., and E. H. Olmsted. *Claude Bernard and the Experimental Method in Medicine*. Henry Schuman, 1952.

Orlans, F. Barbara. *In the Name of Science: Issues in Responsible Animal Experimentation*. Oxford University Press, 1993.

Orlean, Susan. *On Animals*. Avid Reader Press, 2021.

Pernecky, Tomas. *Epistemology and Metaphysics for Qualitative Research*. SAGE, 2016.

Posner, Gerald. *Pharma: Greed, Lies and the Poisoning of America*. Avid Reader Press, 2020.

Pynchon, Thomas. *Against the Day*. Vintage Books, 2006.

Read, John. *From Alchemy to Chemistry*. Dover Books, 1995.

Regan, Tom. *The Case for Animal Rights*. University of California Press, 1983.

Renn, Jurgen. *The Evolution of Knowledge: Rethinking Science for the Anthropocene*. Princeton University Press, 2020.

Riskin, Jessica. *The Restless Clock: A History of the Centuries Old Argument Over What Makes Living Things Tick*. University of Chicago, 2016.

Ritchie, Stuart. *Science Fictions: How Fraud, Bias, Negligence and Hype Undermine the Search for Truth*. Metropolitan Books, 2020.

Roszak, Theodore. *The Making of a Counter Culture: Reflections on the Technocratic Society and Its Youthful Opposition*. Doubleday, 1968.

Rowlands, Mark. *Animals Like Us*. Verso, 2002.

Rubenstein, Richard E. *Aristotle's Children: How Christians, Muslims, and Jews Rediscovered Ancient Wisdom and Illuminated the Middle Ages*. Harcourt, 2003.

Rupke, Nicholas A. *Vivisection in Historical Perspective*. Routledge, 1987.

Ryder, Richard D. *Animal Revolution*. Basil Blackwell, 1989.

Sacks, Oliver. *Hallucinations*. Vintage, 2012.

Safina, Carl. *Beyond Words: What Animals Think and Feel*. Picador, 2015.

Salisbury, Joyce E. *The Beast Within: Animals in the Middle Ages*. Routledge, 1994.

Sarewitz, Daniel. *Frontiers of Illusion: Science, Technology and the Politics of Progress*. Temple University Press, 1996.

Sassi, Maria Michela. *The Beginnings of Philosophy in Greece*. Princeton University Press, 2018.

Shevelow, Kathryn. *For the Love of Animals: The Rise of the Animal Protection Movement*. Holt, 2008.

Singer, Peter. *Animal Liberation*. Ecco, 1975.

Slate, Max. *The Rationalism and Empiricism Debate*. Little Library of Philosophy, 2017.

Sorabji, R. *Animal Minds and Human Morals: The Origins of the Western Debate*. Cornell University Press, 1993.

Sorabji, R. *Aristotle Re-interpreted*. Bloomsbury Academics, 2016.

Svendsen, Lars. *Understanding Animals: Philosophy for Dog and Cat Lovers*. Reaktion Books, 2019.

Szulakowska, Urszula. *Alchemy in Contemporary Art*. Ashgate, 2011.

Taylor, Richard. *Good and Evil: A New Direction*. Prometheus Books, 1984.

Temkin, Owsei. *Galenism: Rise and Decline of a Medical Philosophy*. Cornell University Press, 1973.

Thomas, Keith. *Man and the Natural World: Changing Attitudes in England 1500–1800*. 1983.

Truitt, E. A. *Medieval Robots: Mechanism, Magic and Art*. University of Pennsylvania Press, 2015.

Valenstein, Eliot S. *The War of the Soups and the Sparks: The Discovery of Neurotransmitters and the Dispute Over How Nerves Communicate*. Columbia University Press, 2005.

Weber, Max. *The Protestant Ethic and the Spirit of Capitalism*. George Allen and Unwin, 1930.

Wellmon, Chad. *Organizing Enlightenment: Information Overload and the Invention of the Modern Research University*. John's Hopkins University Press, 2015.

Whitaker, Katie. *Mad Madge: The Extraordinary Life of Margaret Cavendish, Duchess of Newcastle, the First Woman to Live by Her Pen*. Basic Books, 2002.

White, Gilbert. *The Natural History of Selborne* (ed. Anne Secord). Oxford World Classics, 2016.

Wilberding, James, and Christoph Horn. *Neoplatonism and the Philosophy of Nature*. Oxford University Press, 2012.

Wood, Gaby. *Living Dolls*. Faber and Faber, 2002.

Wright, Thomas. *Circulation: William Harvey's Revolutionary Idea*. Chatto and Windus, 2012.

Index

For the benefit of digital users, indexed terms that span two pages (e.g., 52–53) may, on occasion, appear on only one of those pages.

Tables and figures are indicated by *t* and *f* following the page number

absorption, distribution, metabolism, excretion, toxicology (ADMET) testing, 78, 88–89, 91–92
acetaminophen (Tylenol), 73–74
AD (Alzheimer's disease), 101n.36
Addison, Joseph, 198
ADMET (absorption, distribution, metabolism, excretion, toxicology) testing, 78, 88–89, 91–92
adult stem cells (ASCs), 162–63, 166–67
aesthetics, animals and, 110–11, 110*f*
African sleeping sickness, 75
"Against Barbarity to Animals" (Pope), 200
Against the Day (Pynchon), 104–5
ahimsa (nonviolence), 236–37, 239, 240–41, 242
Alcmaeon of Croton, 15
Alexander the Great, 185–86
ALS (amyotrophic lateral sclerosis), 164–65
Also Sprach Zarathustra (Nietzsche), 81
altruism, birds and, 113
Alzheimer's disease (AD), 101n.36
Ambrose (saint), 189
AMCs (arguments from marginal cases), 221–22
American Pet Products Association (APPA), 180
American Society for the Prevention of Cruelty to Animals (ASPCA), 218n.39
amyotrophic lateral sclerosis (ALS), 164–65
anhedonia, 130–31
aniline, 73–74, 75
animals. *See also specific animals*
    aesthetics and, 110–11, 110*f*
    analogies with, 67–68
    Augustine on treatment of, 34
    in cave paintings, 12
    Christianity on humans compared to, 34, 68–69
    consciousness of, 105–6, 115–20
    eating, 10–11
    facial expressions of, 107–9
    fictional stories of, 106–8, 106*f*
    Galen and vivisection of, 24–26
    games played by, 109
    genetics of humans compared to, 82–84, 93–95
    in Golden Age, 12–13
    in "great chain of being," 18–20, 24, 34
    humans as superior to, 12–13, 15, 233
    Kant on human treatment of, 225–26
    language and, 13, 104, 108–9, 192–93
    as machine, 44–46
    Magendie on pain of, 60–61
    metempsychosis and, 15–16
    moral status of, 19–20, 225–27
    nonrationality of, 24
    nonviolence towards, 242
    pain perception of, 118
    as "persons," 232
    self-reflective consciousness of, 128–30
    souls of humans compared to, 18–19, 20
    suffering and, 134–35
    understanding, 104–5
Animal Defense and Antivivisection Society, 212
animal dissection. *See also* vivisection
    Aristotle and, 21–22
    Galen on, 36–37
    by Vesalius, 36–37

animal electricity, Galvani's experiments on, 51–55, 52f
animal experimentation. *See also specific animals; specific methods*
advances from, 224
aim of, 98
animal choice for, 71
in antibiotics discovery, 76
arguments from marginal cases in, 221–22
Aristotle and, 18–22, 26
aversive stimulus and, 1–3
Boyle and, 47–48
on cannabis pharmacology, 90–91
for cardiovascular system of humans, 87–88
challenges of, 6–7
consciousness and, 85
cruelty and, 69–70, 93
development of, 69–70
for drug addiction, 1–3, 2f
Eastern culture and, 236–42
effectiveness of, 80
efforts to change, 182–83
electricity and, 51–53, 52f
ethics of, 222–24, 239–40, 249–51, 268–70
in France, 59–65
frogs for, 50
future of, 270
Galen and, 22–31
Gandhi on, 271
genetic mutations and, 85–86
Harvey on, 38–40
Hippocrates and, 16–18
history of, 11–18
justification for, 222, 227–29
laws for pharmaceutical industry and, 76–79
muscle irritability and, 49–50
nature of, 79–92
off-target effects and, 89–92
as old-fashioned, 173
organoid use instead of, 170
in pharmaceutical industry history, 75–76, 78–79
public opinion on, 181–83
reasons for, 14–15, 18
Sanderson's textbook on, 64
scientific publications and, 229–33
societal changes and, 7
standardization in, 80–81
Stoics on, 20–21
survival surgery and, 63, 93
types of animals used in, 10, 81–84, 81t
usefulness of, 67–68, 88
in vaccine discoveries, 76, 227–29
vacuum pump and, 47–49, 48f
vermin for, 82
Vesalius on, 37–38
Animal Liberation (Singer), 226–27
animal studies, 222–23
animal welfare
in ancient Greece, 187–88
antislavery activism and, 205–6
Brown Dog affair of 1903 and, 210–15, 213f
Cartesianism and, 194–95
in China, 248–49
Enlightenment appreciation for, 194–95
in Great Britain, 207–15
laws on, 208, 209–10, 243–52
Montaigne on, 191–94
as movement, 182–83, 243–44
organizations, 243–44
rise of, 206–15
in Switzerland, 247–48
"3Rs" for, 249–50
in United States, 244–47
Animal Welfare Act (AWA), 244–45
antianxiety drugs (anxiolytics), 140n.58
antibiotics, animal experimentation in discovery of, 76
antidepressants, 132
Antidotum Mithridaticum (mithridatium) formula, 72–73
Antinori, Severino, 160–61
antislavery activism, 205–6
antivivisection movement, rise of, 206–15
anxiolytics (antianxiety drugs), 140n.58
apes, 127
The Apology for Raimond Sebond (Montaigne), 191–92
APPA (American Pet Products Association), 180
applied research, 70–72
arguments from marginal cases (AMCs), 221–22

Aristotle, 12–13, 188
  animal dissection and, 21–22
  animal experimentation and, 18–22, 26
  basic research and, 70
  on "great chain of being," 18–20, 24, 34
  heart and animal experimentation of, 26
  on moral status of animals, 19–20
  on regeneration, 145–46
  on souls of animals compared to
    humans, 18–19
  teleology informing, 18–19
arsenic, 75
arsphenamine (Salvarsan), 75–76,
  95, 130–31
art
  on cruelty, 195–96, 197f
  gerbils and appreciation of, 110–
    11, 110f
  pets inspiring, 186–87, 187f, 200, 201f
  "readymades" by Duchamp, 229–
    31, 230f
Artaud, Antonin, 61–62
Artemesia annua (Sweet Wormwood
  plant), 9–10
artemisinin, 9–10
arteries
  blood of veins compared to, 43
  of dogs, 42
  Fabricius on, 39
  Galen's work on, 26, 28–30, 32n.9
  Harvey's research on, 40–44, 41f
  Herophilus and Erasistratus' work on,
    26–28, 27f
Art Forms in Nature (Haeckel), 150f
ASCs (adult stem cells), 162–63, 166–67
ASPCA (American Society for the
  Prevention of Cruelty to Animals),
  218n.39
assembloids, 171
Atoxyl, 75–76
Augustine (saint), 34–36, 189
automatons, 68–69
aversive stimulus, 1–3
Avicenna, 173
AWA (Animal Welfare Act), 244–45

Bacon, Francis, 59, 68
bananas, genetics and, 83–84
basic research, 70–72

Basset Table (Centlivre), 198
battery, discovery of, 58
Bayer, 73–74
Bayliss, William, 210–12
Beauchamp, Antoine, 75
behavior
  of birds, 113
  in depression, 130–31
  interpreting, 103–4, 113–15
  of octopuses compared to
    humans, 112–13
  of rodents compared to humans, 106–11
behaviorism, 113–15
Bellica Instrumenta (Fontana), 3–4
Benjamin, Walter, 4–5
Bentham, Jeremy, 134, 140n.61, 200–
  2, 226–27
Bentinck, William, 145–46
benzodiazepines, 140n.58
Bergh, Henry, 218n.39
Bernard, Claude, 60–61, 224
  cruelty of, 63
  homeostasis idea of, 62, 95–96
  legacy of, 64
  Magendie working with, 62–64
Bernstein, Julius, 63–64
Bertheim, Alfred, 75–76
Beston, Henry, 117–18
Bhagavad Gita, 237–38
BIA 10-2474, 91–92
Bial, 91–92
biomedical research. See also animal
  experimentation
  arguments from marginal cases
    in, 221–22
  basic and applied, 70–72
  benefits of, 84–85
  birth of modern, 36–44
  clinical, 71–72
  Hippocrates and, 16–18
  killing animals linked to, 10
  knowledge from, 70–71
  NIH funding, 71–72
  preclinical, 76
  repetition in, 59
  Western culture and, 9–10
birds, 113
Blake, William, 202–3, 203f
blastoids, 165

blood, of arteries compared to veins, 43
blood flow
    Fabricius on, 39
    Galen's work on, 26, 28–30, 32n.9
    Harvey's research on, 40–44, 41*f*
    Herophilus and Erasistratus' work on,
        26–28, 27*f*
blood oxygen level dependent (BOLD)
    method, 124–26
bonobos
    genetics of, 83
    self-reflective consciousness test
        on, 128–30
Borges, Jorge Luis, 67–68
Boulle, Pierre, 107
Boveri, Theodor, 151–52, 151*f*
Box, George, 68
Boyle, Robert, 47–48, 261–62
brain. *See also* consciousness
    of birds, 113
    consciousness and electrical activity
        of, 115–16
    DMN and, 124–27
    epilepsy and, 120–21
    Galen's experiments on, 25–26
    ischemic stroke and, 95–96
    live imaging techniques of, 124–26,
        139n.45
    organoids, 167–68
    "split brain" experiments, 137–38n.33
    stroke damage to, 143–44
    visual information and
        damaged, 120–22
    ZIKV and microcephaly of, 170–71
Brainex, 143–44
Brecht, Berthold, 257
Briggs, Robert, 154
British Medical Journal, 214
British Union for the Abolition of
    Vivisection, 213–14
The Brothers Karamazov
    (Dostoevsky), 179
Brown Dog affair, 1903, 210–15, 213*f*
Browne, Thomas, 14, 86–87, 105–
    6, 194–95
Bruno, Giordano, 36–37
Bucephalus the horse, 185–86
Buddhism, 236–39

bullfighting, 10–11
Bulwer-Lytton, Edward, 33
Burch, R. L., 249
burials, of pets, 183–85
Bush, Vannevar, 261–62

CAAT (Center for Alternatives to Animal
    Testing), 251–52
Caenorhabditis elegans (nematode
    worm), 84–85
Caius, John, 38–39
Cambridge Declaration on
    Consciousness, 134
Campbell, Keith, 155–56
cancer, 95
cannabis pharmacology, animal
    experimentation on, 90–91
Canon of Medicine (Avicenna), 173
capitalism, 4–5
cardiovascular disease, Galen's study of
    pulse and, 29
cardiovascular system
    animal experimentation for
        human, 87–88
    evolution of, 86–87
    Fabricius on, 39
    Galen's work on, 26, 28–30, 32n.9
    genetic mutations in, 86
    Harvey's research on, 40–44, 41*f*, 87
    Herophilus and Erasistratus' work on,
        26–28, 27*f*, 32n.9
    of mice, 87–88
    organoids, 167–68
    of snakes, 87–88
    uses of, 87
    Vesalius on anatomy and function of, 38
Carson, Rachel, 226–27
Cartesianism, 194–95
cats, 183–85, 184*f*, 190
Cavendish, Margaret, 195
celebrity pets, 179–80, 185–86
cells. *See also* embryological development
    adult stem, 162–63, 166–67
    historical study of, 146–47
    human embryonic stem, 159, 160–61,
        162–63, 166–67
    induced pluripotent stem, 162–63
    ontogeny and, 147

organoids and, 164–73
pluripotent stem, 162
primordial germ, 159–60
regeneration of, 149
SCNT and, 154–57
stem, 147–48, 151–54, 151*f*
Center for Alternatives to Animal Testing
    (CAAT), 251–52
Centlivre, Susanna, 198
CF (cystic fibrosis), 168
chemotherapy, 74
chimpanzees
    genetics of, 83
    self-reflective consciousness test
        on, 128–30
China, 236, 248–49
cholinesterase inhibitors, 101n.36
Choupette, 179
Christianity
    on humans compared to animals,
        34, 68–69
    on Nature, 188
    rise of, 34
    science impact of, 33–34
    suffering and, 134–35
Churchill, Caryl, 269–70
Cicero, 20–21
cisplatin, 169–70
cis-regulatory elements (CREs), 84
The City of God (Augustine), 34
clinical research, 71–72
Clonaid, 160–61
cloning
    embryological development
        and, 153–54
    "fourteen-day" rule and, 159–60
    hESC and iPSC strategy for, 163
    humans, 158–64
    organoids and, 164–73
    SCNT and, 154–57
    sheep, 155–57, 160–61
    therapeutic, 164–65
Cobbe, Frances Power, 209–10
Coleridge, Stephen, 210–12
Colombo, 32n.9
The Coming Race (Bulwer-Lytton), 33
"comparative physiology," 71
The Confessions (Augustine), 34

consciousness
    animal experimentation and, 85
    of animals, 105–6, 115–20
    Cambridge Declaration on, 134
    challenges of studying, 119–20
    consciousness about, 113–15
    disorders of, 130–34
    DMN and, 124–27
    electrical activity of brain and, 115–16
    evolution of, 133
    Leibniz's Mill argument and, 119–20
    LSD and, 122–24, 127–28, 138n.41
    mechanisms of, 120–22
    meta, 113–15, 126
    mice and, 95–96
    nervous system connected to, 134
    primary, 126
    qualia and, 116–18, 124, 134
    secondary (phenomenal), 126
    self, 124–26, 133
    self-reflective, 128–30
    synesthesia and, 124
    translating, 130–35
    umwelt and, 118–19
Constantine, 34
Controlled Substances Act, 1970, 124
conversion disorder, 105–6
Copernicus, Nicolaus, 36–37, 190–91
corpus callosum, 120–21
Coventry, Francis, 199
COVID-19 pandemic, 180, 227–29
Crane, Walter, 210, 211*f*
CREs (*cis*-regulatory elements), 84
"crisis of reproducibility," science
    and, 264–65
CRISPR/Cas9 gene editing, 84–85, 172
cruelty, 4. *See also* animal welfare
    animal experimentation and, 69–70, 93
    art on, 195–96, 197*f*
    of Bernard, 63
    Cartesianism and, 194–95
    of Galen, 30–31
    of Harvey, 44
    of killing animals, 48–49
    laws on, 208, 209–10, 243–52
    Pope on, 200
    of vivisection, 47
Cruelty to Animals Act, 1849, 208, 210

culture of science, 233–35
cystic fibrosis (CF), 168

Darwin, Charles, 12–13, 133, 205–6
Darwin, Erasmus, 49, 142
death. *See also* killing animals
    defining, 60–61, 143
    in mice studies, 2–3
    in rat studies, 5
    reanimation and, 141–45
Death and Powers, 115–16
De Augmentis Scientiarium (Bacon), 68
Dee, John, 3–4
DeepSqueak, 108–9
default mode network (DMN), 124–27
DEG (diethylene glycol), 77–78
*De humani corporis fabrica (On the Fabric
    of the Human Body)* (Vesalius), 36–37
delta-9-tetrahydrocannabinol
    (THC), 90–91
De Motu Cordis (Harvey), 38–39
depression, 130–34
*De Revolutionibus Orbium Coelestium
    (On the Revolutions of the Heavenly
    Spheres)* (Copernicus), 36–37
Descartes, René, 190–91
    on animals as machine, 44–46
    on human physiology, 45–46
    legacy of, 46
    on Montaigne, 193
desensitization, science and, 233–35
De Theriaca ad Pisonem (Galen), 73
*De viribus electricitatis in motu musculari
    (On the Forces of Electricity in
    Muscular Motion)* (Galvani), 52–53
diabetes, 164–65
diacetylmorphine, 73–74. *See also* heroin
diethylene glycol (DEG), 77–78
digoxin, 72–73
dissecting human corpses, 17–18
dissection. *See* animal dissection
DMN (default mode network), 124–27
DNA, 84–85, 156
dogs
    ancient burials of, 183–84
    arteries of, 42
    Brown Dog affair of 1903 and, 210–
        15, 213*f*

Harvey's vivisection of, 41, 42
    umwelt of, 118–19
Dolly the sheep, 155–57, 160–61
Donne, John, 141, 189
The Doors of Perception (Huxley), 241–42
Dostoevsky, Fyodor, 179
Driesch, Hans, 151–52
drug addiction, animal models for, 1–3, 2*f*
drug development, 49–50
drug metabolism studies, 89
drug off-target effects, 89–92
Dubois-Reymond, Emil, 63–64
Duchamp, Marcel, 229–31, 230*f*
dye-making companies, in early
    pharmaceutical industry, 73–74

Eastern culture, science and, 236–42
eating animals, 10–11
eels, hearts of, 41–42
ego, 126
Ehrlich, Paul, 74–76, 95, 130–31
electrical activity of brain, consciousness
    and, 115–16
electrical capacitors, 51
electricity
    advancements in, 51
    battery discovery and, 58
    Galvani's experiments on muscle
        irritability and, 51–55, 52*f*
    metal, 55–57, 58
    nature of, 53
    Volta's experiments on, 55–57,
        57*f*, 58–59
elephants, 232
E-Life, 264–65
embryological development
    cloning and, 153–54
    "fourteen-day" rule and, 159–60
    frogs and, 151–52
    gastrulation in, 159–60
    Haeckel's work on, 147–49
    salamanders and, 152–54
    stages of, 158–59
    stem cell division and, 151*f*, 151–52
    Weismann on, 150–51
empagliflozin, 169–70
empathy, rats and mice and, 109–10
Empedocles, 185

ENCODE project, 84
endocannabinoids, 90–91
"enframing," 4–5, 6
The Enquiry Concerning Animals
    (Aristotle), 19
environmentalism, 242
epilepsy, 16–17, 120–21
Erasistratus, 23–24
    cardiovascular system work of, 26–28,
        27f, 32n.9
    human vivisection and, 21–22
Erskine, Thomas, 207–8
ethics, of animal experimentation, 222–24,
    239–40, 249–51, 268–70
evolution
    of cardiovascular system, 86–87
    of consciousness, 133
    phylogenetic tree of, 84–85
    tree of, 205–6
An Experiment on a Bird in the Air Pump,
    47–48, 48f
*The Expression of the Emotions in Man and
    Animals* (Darwin, C.), 133

FAAH (fatty acid acyl-hydrolase), 90–92
Fabrizio, Girolamo (Fabricius), 38–39, 42
facial expressions
    of animals, 107–9
    of humans, 107–8
fatty acid acyl-hydrolase (FAAH), 90–92
FDA (Food and Drug Administration),
    77–78, 88–89, 160–61
Fernel, Jean, 45–46
Ficino, Marsilio, 241–42
fictional stories of animals, 106–8, 106f
Fire, Andrew, 84–85
fish
    electricity produced by, 51–52
    hearts of, 41–42
5-HT (serotonin), 112–13, 123–26
5-HT2A receptors, 123–24
fMRI (functional magnetic resonance
    imaging), 124–26, 138n.41
Folkman, Judah, 81–82
Fontana, Giovanni, 3–4
Food, Drug, and Cosmetics Act, 1938, 77–78
Food and Drug Administration (FDA),
    77–78, 88–89, 160–61

"Fountain," 229–31, 230f
The Four Stages of Cruelty, 195–96, 197f
"fourteen-day" rule, embryological
    development and, 159–60
France, physiology in 19th century, 59–65
Frankenstein (Shelley), 142–43, 154–55
Franklin, Benjamin, 51
Franklin square, 51, 52–53, 54–55, 55f
Friedmann, Erika, 181
frogs
    in animal experimentation, 50
    electricity and muscle irritability of,
        51–55, 52f
    embryological development
        and, 151–52
    Leyden jar compared to, 54–55, 55f
    nucleus transfer studies on, 154–55
fruit flies, 86
functional magnetic resonance imaging
    (fMRI), 124–26, 138n.41

Gaia hypothesis, 226–27
Galen, 188
    on animal dissection, 36–37
    animal experimentation and, 22–31
    brain experiments of, 25–26
    cardiovascular system work of, 26, 28–
        30, 32n.9
    competitive nature of, 25
    cruelty of, 30–31
    as gladiator surgeon, 22–23
    Hippocrates influence on, 23–24
    larynx discoveries of, 25–26
    legacy of, 29–31, 33–34
    on nonrationality of animals, 24
    pharmacology experiments of, 73
    pulse and cardiovascular disease
        study of, 29
    Vesalius on errors of, 37–38
    vivisection of animals by, 24–26
    written works of, 23
Gallup, Gordon, 128–30
Galvani, Luigi
    animal electricity experiments of, 51–
        55, 52f
    Harvey's work compared to, 55
    Volta's electricity experiments and, 55–
        57, 57f, 58–59

games, animals playing, 109
Gandhi, Mohandas, 240–42, 271
gastrulation, 159–60
gastruloids, 165
Gegenbaur, Carl, 147
gene editing, CRISPR/Cas9, 84–85, 172
genetics
    of animals compared to humans, 82–
        84, 93–95
    ENCODE project and, 84
    mutations in, 85–86
gerbils, art appreciation and, 110*f*, 110–11
Germany, 63–64
germ theory of disease, 74
Ginsberg, Alan, 240–41
gladiators, Galen as surgeon to, 22–23
Gluck, John P., 6
God, Nature and, 202
Goethe, 203–4, 237–38
Golden Age, animals in, 12–13
Goodall, Jane, 226–27
Great Britain, animal welfare in, 207–15
"great chain of being," 18–20, 24, 34
Greece
    animal welfare in ancient, 187–88
    pet keeping in ancient, 185–88
    science in ancient, 11–14
Grof, Stanislav, 123
Guenon, Rene, 241–42
Guinther, Johannes, 36–37
Gurdon, John, 154–56, 162–63

Haeckel, Ernst, 84, 147–49, 148*f*,
    150*f*, 205–6
Haller, Albrecht, 49–50, 167–68
hallucinogenic drugs, 122–24, 127–28
Handbook for the Physiological Laboratory
    (Sanderson), 64
Happy the elephant, 232
Harvey, William, 224
    on animal experimentation, 38–40
    on blood of arteries compared to
        veins, 43
    cardiovascular system research of, 40–
        44, 41*f*, 87
    cruelty of, 44
    eel and fish heart experiments of, 41–42
    Fabricius and, 39, 42

Galvani's work compared to, 55
    legacy of, 38, 44–45
    on liver and blood production, 43
    research institute of, 40
    vivisection of dogs by, 41, 42
Hata, Sahachiro, 75–76
heart
    Aristotle's animal
        experimentation on, 26
    of eels and fish, 41–42
    Galen's work on, 26, 28–30, 32n.9
    Harvey's research on, 40–44, 41*f*
    Herophilus and Erasistratus' work on,
        26–28, 27*f*
    Vesalius on anatomy and function
        of, 38
Heidegger, Martin, 4–5
"Herbert West—Reanimator"
    (Lovecraft), 142–43
Hermann, Ludimar, 63–64
heroin, 1–2, 73–74, 89
Herophilus, 23–24
    cardiovascular system work of, 26–28,
        27*f*, 32n.9
    human vivisection and, 21–22
Herschel, John, 67
Hertwig, Oscar, 151–52
Hertwig, Richard, 151–52
hESCs (human embryonic stem cells),
    159, 160–61, 162–63, 166–67
Hesiod, 12–13, 145
Hillel (rabbi), 221
Hinduism, 44–45
Hippocrates, 16–18, 23–24
*The History of Pompey the Little or the
    Life and Adventures of a Lap-Dog*
    (Coventry), 199
HIV-1 virus, mice studies on, 95
Hobbes, Thomas, 103
Hodgson, J., 88–89
Hofmann-LaRoche, 140n.58
Hogarth, William, 195–96, 197*f*
Holocaust, 226–27
homeostasis, Bernard and idea of,
    62, 95–96
L'homme machine (Man the Machine)
    (La Mettrie), 49
Hooke, Robert, 47–48, 59–60, 146–47

human embryonic stem cells (hESCs),
    159, 160–61, 162–63, 166–67
humanities, research universities
    and, 266–67
humans
    animal experimentation for
        cardiovascular system of, 87–88
    animals as inferior to, 12–13, 15, 233
    Christianity on animals compared to,
        34, 68–69
    cloning, 158–64
    depression in, 130–31
    dissecting corpses of, 17–18
    eating animals, 10–11
    experimentation on, 78–79
    facial expressions of, 107–8
    genetics of animals compared to, 82–
        84, 93–95
    in "great chain of being," 18–20, 24, 34
    Kant on animal treatment by, 225–26
    metempsychosis and, 15–16
    mice compared to, 94, 111
    octopus behavior compared to, 112–13
    physiology of, 45–46
    reasons for not experimenting
        on, 16–18
    rodent behavior compared to, 106–11
    souls of animals compared to, 18–19, 20
    uses for killing animals, 10–11
    vivisection, 21–22
Humboldt, Wilhelm von, 259–60
Hume, David, 194–95
hunting
    Cavendish on, 195
    laws on, 209–10
"The Hunting of the Hare"
    (Cavendish), 195
Huxley, Aldous, 241–42
Huysman, J.-K., 4

IACUCs (Institutional Animal Care and
    Use Committees), 221, 246–47
ICM (inner cell mass), 159, 160–61
India, 236–37
induced pluripotent stem cells
    (iPSCs), 162–63
Industrial Revolution, 207
inner cell mass (ICM), 159, 160–61

Institutional Animal Care and Use
    Committees (IACUCs), 221, 246–47
internal review boards (IRBs), 226–27
interoception, 115–16
intestine organoids, 169
Introduction to the Principles of Morals
    and Legislation (Bentham),
    140n.61, 200–1
Introduction to the Study of Experimental
    Medicine (Bernard), 64
iPSCs (induced pluripotent stem
    cells), 162–63
IRBs (internal review boards), 226–27
ischemic stroke, 95–96
Ishiguro, Kazuo, 161, 269–70
Islam, science impact of, 33–34
The Island of Dr. Moreau (Wells), 154–55
"isolated tissue preparation," 49–50

Jainism, 236–37
James, Henry, 104–5
Johns Hopkins University, 251–52
Johnson, Samuel, 200
Jones, William, 237–38
Journal of the National Antivivisection
    Society, 210, 211f

Kant, Immanuel, 200–1, 225–26
kidney organoids, 169–70
killing animals. See also death; vivisection
    biomedical research linked to, 10
    cruelty of, 48–49
    decisions on, 234–35
    desensitization and, 235
    human interests in, 10–11
    indifference to, 6
    mice compared to rats, 6
    process of, 5
    terminology in, 6
King, Martin Luther, Jr., 240–42
King, Thomas, 154
King Lear (Shakespeare), 67
Klein, Emmanuel, 209–10
Koch, Robert, 74

Lagerfeld, Karl, 179
La Mettrie, Julien Offray de, 49
Lancet, 265–66

language, animals and, 13, 104, 108–9, 192–93
larynx, Galen's discoveries on, 25–26
laws
  on animal welfare and cruelty, 208, 209–10, 243–52
  on hunting, 209–10
  on IRBs, 226–27
  pharmaceutical industry and, 76–78, 92
  against vivisection, 209–10
L-DOPA, 101n.36
learned helplessness, in mice, 131–32
Leary, Timothy, 123
Le Cat, Claude-Nicolas, 68–69, 82
Legallois, Julien, 60–61
Leibniz, Gottfried, 119–20
Leibniz's Mill argument, 119–20
Leviathan (Hobbes), 103
Levi-Strauss, Claude, 12
Leyden jar, 51, 52–53, 54–55, 55f
life
  medicine devoted to, 144
  reanimation and, 141–45
  suffering and will to, 238
  systems of, 142
Light and Colour (Goethe's Theory)—The Morning after the Deluge—Moses Writing the Book of Genesis, 204f
Lind af Hageby, Lizzy, 210–12
live brain imaging techniques, 124–26, 139n.45
liver
  blood production by, 43
  regeneration, 145–46
Locke, John, 194, 217n.24
Lovecraft, H. P., 142–43
LSD, consciousness and, 122–24, 127–28, 138n.41
Lunar Society, 49

macular degeneration, 164–65
Magendie, Francois
  on animal pain, 60–61
  Bernard working with, 62–64
  cruelty of, 60–62
Magia Universalis (Schott), 3–4
Magnus, Albertus, 189

Maharishi Mahesh Yogi, 240–41
Majer, Friedrich, 237–38
Malebranche, Nicolas, 46
Malpighi, Marcello, 43–44
Mann, Thomas, 3–4
Man the Machine (L'homme machine) (La Mettrie), 49
Mario and the Magician (Mann), 3–4
Martin, Richard, 208
MDMA (methylenedioxymethamphetamine), 112–13
medicinal chemistry, 76
Mello, Craig, 84–85
mentors, at research universities, 267–68
meprobamate, 140n.58
meta-consciousness, 113–15, 126
metal electricity, 55–57, 58
metempsychosis, 15–16
methylenedioxymethamphetamine (MDMA), 112–13
mice
  cardiovascular system of, 87–88
  consciousness and, 95–96
  death in studies of, 2–3
  depression and, 131–32, 133–34
  DMN in, 127
  drug addiction and, 1–3, 2f
  empathy and, 109–10
  facial expressions of, 107–9
  fictional stories of, 106–8, 106f
  genetics and, 83, 93–95
  hallucinogenic drugs and, 127–28
  HIV-1 virus in studies on, 95
  humans compared to, 94, 111
  ischemic stroke and, 95–96
  language and, 108–9
  learned helplessness in, 131–32
  mind of, 105–7
  qualia of, 134
  rat studies compared to, 6
  standardization in studies of, 80–81
  success of studies on, 81–82
  usefulness of studies on, 88, 92, 94
microcephaly, 170–71
microfluidics, 171
microscopy, 145–46
Mill, John Stuart, 226–27
Mithridates VI, 72–73

mithridatium *(Antidotum Mithridaticum)* formula, 72–73
Monroe, Marilyn, 78
Montaigne, Michel de, 191–94
moral status of animals, 19–20, 225–27
Morgan, Lloyd, 113–15
morphine, 72–74, 89
"Mrs. Winslow's Soothing Syrup," 76–77
Muffett, Thomas, 194–95
Muller, Johannes, 63–64, 146–47
muscle irritability
    electricity and, 51–55, 52*f*
    Haller's work on, 49–50
mutations, genetic, 85–86
"Myth of Er," 15–16

Nagel, Thomas, 116–18
Nanog, 162–63
National Anti-Vivisection Society, 209–10, 213–14
National Institutes of Health (NIH), 71–72, 246
natural history, 14
The Natural History of Selbourne (White), 14
Nature
    Christianity on, 188
    Enlightenment appreciation for, 194–95
    God and, 202
    pharmacology and, 73
    science and, 14, 70
Nature, 84, 94–95, 143–44, 172
Nature Neuroscience, 108–9
Nazis, 226–27
Neanderthals, 172
nematode worm *(Caenorhabditis elegans)*, 84–85
nervous system, consciousness connected to, 134
neuroimmunology, 105–6
Never Let Me Go (Ishiguro), 161, 269–70
Newcomb, Simon, 261–62
Newton, 203*f*
Newton, Isaac, 179–80, 202–4
Nietzsche, Friedrich, 81, 141–42, 250–51
NIH (National Institutes of Health), 71–72, 246
NIH Revitalization Act, 246, 249–50

Nixon, Richard, 124
nonviolence *(ahimsa)*, 236–37, 239, 240–41, 242
NOVA1 gene, 172
A Number (Churchill), 269–70

octopus, 112–13
"off label" drugs, 97
off-target effects, 89–92
On Abstinence from Animal Food (Porphyry), 20–21
On Anatomical Procedures (Galen), 36–37
On Cruelty (Montaigne), 191–92
On Exactitude in Science (Borges), 67–68
On Piety (Theophrastus), 20
On the Cleverness of Animals (Plutarch), 20–21
On the Fabric of the Human Body (De humani corporis fabrica) (Vesalius), 36–37
On the Forces of Electricity in Muscular Motion (De viribus electricitatis in motu musculari) (Galvani), 52–53
On the Generation of Animals (Aristotle), 18–19
On the Parts of Animals (Aristotle), 19
On the Sacred Disease (Hippocrates), 16–17
On the Soul (Aristotle), 18–19
ontogeny, 147
opioid drugs, 89
organic chemistry, 73
organogenesis, 159–60
organoids, cloning and, 164–73
"organ on a chip" systems, 171
Orlean, Susan, 106–7
Ottoman Turks, 34

pain, animal perception of, 118
Paracelsus, 74, 99n.8
Parkinson's disease (PD), 101n.36, 164–65
Pasteur, Louis, 74
Pax (Aristophanes), 3–4
PD (Parkinson's disease), 101n.36, 164–65
People for the Ethical Treatment of Animals (PETA), 243–44
The Perennial Philosophy (Huxley), 241–42
Peritas the dog, 185–86

PETA (People for the Ethical Treatment of Animals), 243–44
pets
    ancient, 183–88
    art inspired by, 186–87, 187f, 200, 201f
    burials of, 183–85
    celebrity, 179–80, 185–86
    diet of, 180–81
    health benefits of owning, 180–81
    labor and utility of, 190
    in Middle Ages, 188–91
    spending on, 180
    valuation of, 190
    witchcraft and, 189–90
PGCs (primordial germ cells), 159–60
Pharma (Posner), 97
pharmaceutical industry
    ADMET testing in, 78, 88–89, 91–92
    animal experimentation's history in, 75–76, 78–79
    drug metabolism studies and, 89
    dye-making companies in early, 73–74
    germ theory of disease and, 74
    laws and, 76–78, 92
    "off label" drugs and, 97
    off-target effects and, 89–92
    organoids for, 168–70
    rise of, 72–79
    syphilis treatment and, 74–76
pharmacology
    of cannabis, 90–91
    Galen's experiments on, 73
    history of, 72
    Nature and, 73
phenomenal consciousness (secondary consciousness), 126
Philosophical Transactions of the Royal Society, 196–98
phylogenetic tree of evolution, 84–85
physiology
    "comparative," 71
    human, 45–46
    in 19th century France, 59–65
pigs, 143–44
Planck, Max, 252
Plato, 15–16, 18
pluripotent stem cells (PSCs), 162
Plutarch, 20–21

"pneuma," 26–28
Pneumatik Engine (vacuum pump), 47–49, 48f
Poe, Edgar Allen, 205
Poison Squad, USDA, 76–77
The Politics (Aristotle), 18–19
Pope, Alexander, 82, 199–200
Porphyry, 20–21, 185, 186
Posner, Gerald, 97
Potter, Beatrix, 106–8, 113–15
preclinical research, 76
primary consciousness, 126
primordial germ cells (PGCs), 159–60
Princess Casamassima (James), 104–5
The Principles of Humane Experimental Technique (Russell and Burch), 249
Protection of Animals Act, 1911, 208
PSCs (pluripotent stem cells), 162
psychedelic drugs, 122–24, 127–28
publications, scientific, 229–33
public opinion, on animal experimentation, 181–83
pulse, Galen's study of cardiovascular disease and, 29
Pynchon, Thomas, 104–5
Pyrrho of Elis, 191
Pythagoras, 15–16, 185

qualia, consciousness and, 116–18, 124, 134
quinine, 72–73

Rachels, James, 205–6
Raëlians, 160–61
rats
    death in studies on, 5
    empathy and, 109–10
    games played by, 109
    mice studies compared to, 6
"readymades," by Duchamp, 229–31, 230f
reanimation, 141–45
Regan, Tom, 232–33
regeneration. See also cloning
    Aristotle on, 145–46
    of cells, 149
    in history, 145–58
    liver, 145–46

mythology on, 145–46
organoids and, 164–73
reanimation and, 141–45
Regiomontanus, 3–4
Remak, Robert, 63–64, 146–47
repetition, in science, 59
Republic (Plato), 15–16
research ethics, 249–50
research universities
    "crisis of reproducibility" and, 264–65
    ethics debated at, 268–70
    financial demands of, 258, 262–63
    future of, 267–71
    historical evolution of, 258–60
    humanities and, 266–67
    mentors at, 267–68
    modern, 257, 259–60
    post-WWII expansion of, 260–63
    systemic problems facing, 264–66
    training at, 263–64
Reynolds, Joshua, 200
The Rise and Fall of the City of Mahagonny
    (Weil and Brecht), 257
RNA, 84–85
RNA interference, 84–85
rodents, behavior of humans compared to,
    106–11. *See also* mice; rats
Rollin, Bernard, 222–23
Roman Empire, 34
Romanticism, 203–6
Romulus Augustulus, 34
Rousseau, Jean-Jacques, 205
Roux, Wilhelm, 151–52
Royal Society for the Prevention of Cruelty
    to Animals (RSPCA), 208–10
Russell, W. M. S., 249

salamanders, 152–54
Salerno, 259
Salvarsan (arsphenamine), 75–76,
    95, 130–31
Sanderson, John Burdon, 64
SARS-CoV-2, 227–29
Schartau, Leisa, 210–12
schizophrenia, 132
Schopenhauer, Arthur, 201–2, 237–39
Schott, Gaspar, 3–4
Schwann, Theodor, 63–64, 146–47

science. *See also* animal experimentation;
    biomedical research
    in ancient Greece, 11–14
    arguments from marginal cases
        in, 221–22
    Augustine on, 35–36
    battery discovery and, 58
    Christianity and Islam's impact
        on, 33–34
    competition and progress in, 58–59
    "crisis of reproducibility" and, 264–65
    culture of, 233–35
    definition of, 11–12
    desensitization and, 233–35
    Eastern culture and, 236–42
    Hippocrates and, 16–18
    history of, 11–18
    middle class interest in, 48–49, 59–
        60, 196–98
    Nature and, 14, 70
    as profession, 59–60
    repetition in, 59
Science, 107–8, 109, 113
scientific publications, 229–33
Scientific Revolution, 33–34, 47–
    48, 190–91
SCNT (somatic cell nuclear
    transfer), 154–57
sea urchins, 151–52
secondary consciousness (phenomenal
    consciousness), 126
Seed, Richard, 160–61
selective serotonin reuptake inhibitors
    (SSRIs), 123–24
self-consciousness, 124–26, 133
self-reflective consciousness, 128–30
SE Massengill, 77–78
serotonin (5-HT), 112–13, 123–26
serotonin transporter (SERT), 112–13
Shakespeare, William, 9, 67
The Shambles of Science (Lind af Hageby
    and Schartau), 210–12
sheep, 155–57, 160–61
Shelley, Mary, 62, 142–43, 154–55
Sheridan, Richard Brinsley, 207
sickle cell disease, 84–85
Silent Spring (Carson), 226–27
Singer, Peter, 226–27

slavery, 205–6
snakes, 18–19, 71
　cardiovascular system of, 87–88
Snow, C. P., 267
Society for the Prevention of Cruelty to
　Animals (SPCA), 208–9
somatic cell nuclear transfer
　(SCNT), 154–57
Some Thoughts Concerning Education and
　Cruelty (Locke), 194
souls, of animals compared to humans,
　18–19, 20
Spallanzani, Lazzaro, 146
SPCA (Society for the Prevention of
　Cruelty to Animals), 208–9
Spemann, Hans, 152–54
Spence, Joseph, 199
Sperry, Roger, 120
Spinoza, Baruch, 202
"split brain" experiments, 137–38n.33
SSRIs (selective serotonin reuptake
　inhibitors), 123–24
Stammbaum (stem tree), 147, 148f
standardized animal models, 80–81
Starling, Ernest, 210–12
stem cells, 147–48, 151–54, 151f
stem tree (Stammbaum), 147, 148f
Stoics, on animal experimentation, 20–21
Streisand, Barbara, 179–80
stroke, 95–96
　brain damage from, 143–44
suffering. See also cruelty
　animals and, 134–35
　Bentham on, 200–1
　Christianity and, 134–35
　will to life and, 238
survival surgery, 63, 93
Sweet Wormwood plant (Artemesia
　annua), 9–10
Swift, Jonathan, 107, 137–38n.33, 194–
　95, 198
Switzerland, animal welfare law in, 247–48
synesthesia, consciousness and, 124
syphilis, 74–76

teleology, 18–19
THC (delta-9-tetrahydrocannabinol),
　90–91

Theodosius, 34
Theogony (Hesiod), 145
Theophrastus, 20
therapeutic cloning, 164–65
Thomas, Keith, 189
Thomas Aquinas (saint), 34, 189
"3Rs" for animal welfare, 249–50
three-dimensional printing, 171
Thus Spoke Zarathustra
　(Nietzsche), 141–42
TLR3 (toll-like receptor 3), 170–71
toll-like receptor 3 (TLR3), 170–71
torture, 4
"To Science" (Poe), 205
Trembley, Abraham, 145–46
Treponema pallidum, 75, 130–31
Turner, J. M. W., 204f, 205–6
Tu Youyou, 9–10
Twelfth Night (Shakespeare), 9
Tylenol (acetaminophen), 73–74
type 1 diabetes, 164–65

Uexküll, Jakob von, 118–19
umwelt, consciousness and, 118–19
United States, animal welfare law
　in, 244–47
universities. See research universities
University of Berlin, 259–60
Upanishads, 237–38
urodeles, 146
U.S. Department of Agriculture (USDA),
　76–77, 244–45

vaccines, animal experimentation in
　discovery of, 76, 227–29
vacuum pump (Pneumatik Engine), 47–
　49, 48f
Vane, John, 49–50
veins, blood of arteries compared to, 43
vermin, for animal experimentation, 82
Vesalius, Andreas, 36–38
Victoria Street Society, 209–10
Virchow, Rudolf, 63–64, 146–47
vivisection
　antivivisection movement, 206–15
　Brown Dog affair of 1903 and, 210–
　15, 213f
　cruelty of, 47

of dogs by Harvey, 41, 42
eighteenth-century criticism of, 198–99
Galen and animal, 24–26
by Haller, 49–50
human, 21–22
justifications for, 198
laws against, 209–10
Volta, Alessandro, 55–57, 57f, 58–59
"voltaic pile," 58

Wagner, Richard, 7, 239–40
Walker, Karen, 155–56
Weil, Kurt, 257
Weismann, August, 150–51
Wellmon, Chad, 233–34
Wells, H. G., 154–55
Western culture, biomedical research
    and, 9–10

"What Is It Like to Be a Bat?"
    (Nagel), 116
White, Gilbert, 13–14, 194–95
Wilberforce, William, 205–6, 207
Wiley, Harvey W., 76–77
will to life, suffering and, 238
Wilmut, Ian, 155–56
witchcraft, pets and, 189–90
Wittgenstein, Ludwig, 68, 72, 104, 113–15
Wright, Joseph, 47–48, 48f, 196–98

Yamanaka, Shinya, 162–63
yeast cells, 84

Zaks, Tal, 227–29
Zavos, Panayiotis, 160–61
Zika virus (ZIKV), 170–71
Zimietzki, Friedrich W., 233–34